Disasters, Vulnerability, and Narratives

This book uses narrative responses to the 2010 Haiti earthquake as a starting point for an analysis of notions of disaster, vulnerability, reconstruction and recovery. The turn to a wide range of literary works enables a composite comparative analysis, which encompasses the social, political and individual dimensions of the earthquake.

This book focuses on a vision of an open-ended future, otherwise than as a threat or fear. Mika turns to concepts of hinged chronologies, slow healing and remnant dwelling. Weaving theory with attentive close-readings, the book offers an open-ended framework for conceptualising post-disaster recovery and healing. These processes happen at different times and must entail the elimination of compound vulnerabilities that created the disaster in the first place. Challenging characterisations of the region as a continuous catastrophe this book works towards a bold vision of Haiti's and the Caribbean's futures.

The study shows how narratives can extend some of the key concepts within discipline-bound approaches to disasters, while making an important contribution to the interface between disaster studies, postcolonial ecocriticism and Haitian Studies.

Kasia Mika is a Postdoctoral Researcher in Comparative Caribbean Studies at the Royal Netherlands Institute of Southeast Asian and Caribbean Studies and teaches at the University of Amsterdam. Her other work is published, among others, in *The Journal of Haitian Studies*, *Moving Worlds: A Journal of Transcultural Writing*.

Routledge Studies in Hazards, Disaster Risk and Climate Change

Series Editor: Ilan Kelman

Reader in Risk, Resilience and Global Health at the Institute for Risk and Disaster Reduction and the Institute for Global Health, University College London (UCL)

This series provides a forum for original and vibrant research. It offers contributions from each of these communities as well as innovative titles that examine the links between hazards, disasters and climate change, to bring these schools of thought closer together. This series promotes interdisciplinary scholarly work that is empirically and theoretically informed, with titles reflecting the wealth of research being undertaken in these diverse and exciting fields.

The Institutionalisation of Disaster Risk Reduction
South Africa and Neoliberal Governmentality
Gideon van Riet

Understanding Climate Change through Gender Relations
Edited by Susan Buckingham and Virginie Le Masson

Climate Change and Urban Settlements
Mahendra Sethi

Community Engagement in Post-Disaster Recovery
Edited by Graham Marsh, Iftekhar Ahmed, Martin Mulligan, Jenny Donovan and Steve Barton

Climate, Environmental Hazards and Migration in Bangladesh
Max Martin

Governance of Risk, Hazards and Disasters
Trends in Theory and Practice
Edited by Giuseppe Forino, Sara Bonati and Lina Maria Calandra

Disasters, Vulnerability, and Narratives
Writing Haiti's Futures
Kasia Mika

For more information about this series, please visit: www.routledge.com/Routledge-Studies-in-Hazards-Disaster-Risk-and-Climate-Change/book-series/HDC

Disasters, Vulnerability, and Narratives

Writing Haiti's Futures

Kasia Mika

Routledge
Taylor & Francis Group

LONDON AND NEW YORK

First published 2019
by Routledge
2 Park Square, Milton Park, Abingdon, Oxon OX14 4RN

and by Routledge
52 Vanderbilt Avenue, New York, NY 10017, USA

First issued in paperback 2020

Routledge is an imprint of the Taylor & Francis Group, an informa business

British Library Cataloguing-in-Publication Data
A catalogue record for this book is available from the British Library

Library of Congress Cataloging-in-Publication Data
A catalog record has been requested for this book

ISBN 13: 978-0-367-58849-6 (pbk)
ISBN 13: 978-1-138-30075-0 (hbk)

Typeset in Times New Roman
by Wearset Ltd, Boldon, Tyne and Wear

Contents

Acknowledgements

This book is the fruit of countless displacements, re-routings and exchanges with the many teachers and friends I have been lucky enough to encounter along the way. It is my hope that they will see *Disasters, Vulnerability, and Narratives: Writing Haiti's Futures* as a way of showing my gratitude and thanks for their generosity and unwavering support. The book started out as a thesis on narrative responses to the 2010 Haiti earthquake and would not have been possible without the careful and meticulous work of my supervisor at the University of Leeds, Prof Graham Huggan. For his enthusiasm, support and patience with my writing and ever-expanding footnotes, I thank him. I am particularly grateful to him for helping me think through the methodological coordinates of the thesis and for allowing my numerous displacements to shape the project. Rigorous feedback from Charles Forsdick and Samuel Durrant, my examiners and my mentors over the years, helped to further nuance my argument. I also thank friends and colleagues in the School of English at the University of Leeds: Raghini Mohite, Emma Trott, Jay Parker, Ryan Topper and Dominic O'Key. The thesis and this book would not have been possible without the support of the University of Leeds Research Scholarships (2012–15). I thank the University of Leeds for having faith in my work. The PhD started with an order. Special thanks to Adrienne Janus for setting the challenge and for her faith in me during my time at the University of Aberdeen. Nick Nesbitt brought Haiti's contemporary politics into focus, years ago, and introduced me to the work of Edwidge Danticat. The closing words of *Brother, I Am Dying* were the beginning of a life-long journey for me. Janet Stewart always found time and words of advice, both academic and emotional. Simon Ward and Shane Alcobia-Murphy nurtured my curiosity and curbed my stylistic carelessness.

I am very grateful to late Anthony Carrigan for his collegial support, generosity and the many conversations on disasters, literature and Haiti we had shared over the years. His work and words continue to echo across these pages. I wish he could read this book as well as see the coming together of the documentary on creativity and Haiti that we had hoped to make together. Despite all, the film did come together. Ed Owles (Postcode Films) brought a camera, a new perspective, and we spent hours talking, interviewing, filming and waiting together in Haiti. The conversations we shared and the shots we filmed gave a new title and added a different layer of commitment and relevance to this book. For this generous, creative detour I want to thank him, Clare Barker, Matt Boswell and the AHRC (Arts and Humanities Research Council, UK). Hours of multilingual, cross-country and transatlantic conversations with Basile Despland, evenings of poetry, art, performance and nights of *kompa* and pulsing creativity reverberate throughout the film and the pages of this book. For their defiant energy that creates new words and thrusts us into more daring visions of the world and its futures, I want to thank: James Noël, Pascale Monnin, Mario Benjamin, Yanick Lahens, Maksaens Denis, Patrick Vilaire, Princes Eud and all those *enragés* who shared their studios, their plans and their dreams. These journeys of inquiry continued and flourished during my first postdoc position in Comparative Caribbean Studies at KITLV (Royal Netherlands Institute for Southeast Asian and Caribbean Studies). I could not have hoped for a more supportive and inquisitive academic community. I am particularly grateful to Rosemarijn Hoefte, for her ongoing mentorship, meticulous comments and unwavering sense of humour. Henk Schulte Nordholt and Gert Oostindie for having faith in my work and trust in the detours it took. Rivke Jaffe for her example, encouragement and scholarly hospitality. Joeri Arion brought coffee-flavoured laughter, Malcom Ferdinand undisciplined creativity and a pan-Caribbean, reverberating vision of the world into the project. At the University of Amsterdam, I want to thank Jeff Diamanti, David Duindam and Boris Noordenbos for their life-saving irony and support, at times when there were not enough hours in the day.

Across the many locations of this book, I am grateful to Ward Berenschot, Paul Bijl, Charlotte Hammond, Philip Kaisary, Gerry van Klinken, Roland Remenyi, Jessica Vance Roitman, Caroline Sanderson, Michał Szczepański, Marianne Vysma and Wouter Veenendaal who gave their time and provided invaluable support and feedback across the many stages of the thesis, the book proposal and, finally, this manuscript. I am especially grateful to Alessandra Benedicty-Kokken for her inspiring intellectual rigour and care in the most turbulent, seemingly unending, final stages of the book project. Special thanks must be made to staff and friends at the Haitian Creole Language and Culture Institute, University of Massachusetts, Boston, the Institute for Justice and Democracy in Haiti and Haiti Support Group for helping me see the relevance of my work beyond the confines of the academic world. Jean-Rene Lesley, in particular, has helped my vocabulary increase and my vision expand. *Djanm!* Diasporic friendships forged and sustained this project. Aleksander Adamski, Maria Buenaventura, Aleksandra Bujwan, Peter Kravos, Robbie Kubala, Agata Lipczik, Jolanta Lundgren, Wayne Modest, Eunice Ong, Jean-Philippe Mackay, Ania Pomierny-Wąsińska, Idoya Puig, Sandhya Rasquinha, Marcos Silveira, Francis Thomas and Elizabeth Watkins accompanied me at different stops of the journey bearing my irritation, teary fatigue and academic restlessness. Finally, Kevin De Sabbata never grew tired of trying to keep track of where I was, where I was going and of being there when I was finally back. I will be forever grateful for their encouragement, example and loving care.

Sections of Part II were originally published in the *Journal of Haitian Studies*, 20.2 (Fall 2014) under the title 'Histories of the Past, Histories for the Future: Representing the Past and Writing for the Future in Rodney Saint-Éloi's *Haïti, kenbe la!'* Sections of Part III appeared earlier, in significantly different form, in *Moving Worlds: Catastrophe & Environment*, 14.2 (2014), 93–107 under the title 'Recovery Foreclosed: History, Landscape, and Personal: Intervention in Sandra Marquez Stathis's *Rubble: The Search for a Haitian Boy'*. This book would not have been possible without the superb scholarly and editorial guidance at Routledge. I would like to thank Ilan Kelman for his academic mentorship from our first heated debates on disasters and solutions to the publication day of this book. For Ruth Anderson and Faye Leerink, I cannot say enough for their editorial precision, patience and expertise.

My parents, Jolanta and Krzysztof, and my family engendered in me an unquenched curiosity and love for books and far-away destinations. The sacrifices they made are countless and so are the lessons they have taught me over the years. My late grandad, Prof Stanisław Mika, waited and waited to see the completion of this project, reading just about anything I ever wrote, and always hoping to hear more, read more and know more of whatever I was doing. His scholarly inquisitiveness will remain an example for us all. A special thanks to Józek who made me smile even at the most turbulent of times. Arthur Rose has been a more patient and caring friend and interlocutor than I can ever thank him for, talking through just about every idea and anxiety that arose over the course and across the many scattered locations of this project. This book would not have been possible without their unconditional support, and the faith and hope they inspired in me.

Part I
Disaster making

1   12 January 2010

From hazard to disaster

The 12 January 2010 earthquake in Haiti 'caused' the loss of hundreds of thousands of human lives and futures. There was no right place or time when the first 35-second long shock, described by one survivor as lasting 'less than a minute',[1] hit and rocked the island. The earthquake happened at 4.53 pm local time (21.53 UT) approximately 25 km WSW of the capital city of Port-au-Prince (18.443°N, 72.571°W).[2] The magnitude Mw7 tremors, with a depth of 13 km,[3] became a disaster. Over 80 per cent of Port-au-Prince and the nearby Léogâne were levelled, leaving behind anything between 19–40 million m³ of debris.[4] The tremors destroyed around 80 per cent of schools in Port-au-Prince, 60 per cent of government buildings located in the capital as well as 60 per cent of schools in Haiti's South and West Departments.[5] The estimated number of casualties, an uncertainty which might never be resolved, ranges from 100,000 to 300,000.[6] In the days following the disaster, cries of grief and stories of loss and pain joined expressions of joy and relief and came together in a distinct urban chorus, filling the void of the ravaged city and countering the 'executive silence' of the Haitian government.[7] The tremors changed everything, thrusting the country into a new time, one of an ongoing aftermath. Similarly to the redefined landscape of the capital, the lives of those affected by the disaster, whether living in Haiti or elsewhere, were, from now on, rooted in and re-routed by the echoing roar of *gou-dougoudou*, an onomatopeia for the earthquake.[8] Anchored in narratives of the January tremors, this book centres its inquiry on disaster ontologies. It questions the very notions of what disasters are, interrogates the imbricated layers of causes and structures they reveal, attends to the types of relations and they form and representations they call forth. Or, in Veena Das and Arthur Kleinman's words: 'The choke and sting of experience only becomes real – is heard – when it is narrativized.'[9]

Disaster accounts – such as the seven texts by Rodney Saint-Éloi, Lionel-Édouard Martin, Dany Laferrière, Sandra Marquez Stathis, Nick Lake, Dan Woolley and, and Laura Wagner to which I turn throughout *Disasters, Vulnerability, and Narratives: Writing Haiti's Futures* – give voice to such stings of experience, reflect the global nature of the January earthquake and creative responses to it, capture the varied and oft discordant memories of the event, its personal, collective and global meanings and translations. The narratives also help to understand what disasters and long-term vulnerability are and how they are experienced by individuals and communities. These selected texts are characterised by their networked and enmeshed positioning which goes beyond and problematises easy and binary categorisations of insider/outsider, testifying, at the same time, to the multiple, simultaneous points of reference, coordinates, from which these authors write and around which their narratives are organised. Together, the subsequent chapters all work towards advancing these new visions of Haiti and nuanced definitions of disaster, vulnerability, reconstruction, remaking and recovery that are attentive to the long-term processes and histories of unequal power relations that shaped the Caribbean. These categories are rooted in specific understandings of time, space and self which, in turn, are a point of entry into wider examination of the three-way relationship between complex disasters, aesthetic and narrative conventions, and the wider conceptual frameworks used to explain the event's history and its individual and collective impact.

As expressions of various forms of enmeshed and polyvalent witnessing to the disaster and its aftermath, as well as one's attachment to Haiti, these narratives stand at a *crossroads* (*kalfou*/*kafou* in Haitian Kreyòl) – caught between wor(l)ds, epistemologies, imaginaries and pointing to new relations and necessary crossings of methodologies.[10] These narratives witness to loss, provide a testimony for such a traumatic event, are attempts to give and negotiate its meaning, while offering a way to reveal and rework the cultural framework through which disasters, their past, present and future, are perceived.[11] In effect, these selected narratives share in their effort to bear the weight and bear the record of the disaster while also forming an attempt to take one's bearings, that is 'to determine one's position' (*OED*),[12] with regards to the radically changed, and oft unrecognisable, surroundings. Moreover, by engaging with material contexts as well as the collective and individual meanings of the disaster, distinct ways of making sense of its past, present and future, these narratives of the earthquake draw out the many histories of the event, demonstrating, at the same time, how overlapping factors – ones that extend well beyond Haiti's borders or categories of the nation-state, and include but are not limited to issues of history, politics and ecology – determined the scale of the 2010 disaster and shaped the immediate responses to it. In the process of doing so, these texts call for processual definitions of disaster and ruination, call forth differentiated notions of remaking, rebuilding and reconstruction, and try to envisage non-linear and non-teleological recovery.

The January 2010 disaster is a process, not a closed-off event, with *hinged chronologies* and pre-disaster multi-scalar vulnerability defining its scale and the ways in which it continues to unfold into the future. Similarly, post-disaster ruin and ruination extend beyond the immediate post-earthquake destruction. For one, these categories can refer to specific depictions of devastated urban landscapes; and, for another, their temporal signification may connote either a sense of permanence and impenetrability or point to the long-term processes that contributed to this state of ruination. Consequently, recovery emerges as a state of *remnant dwelling* among and beyond the ruins; that is a dual attempt to remake and conceive of one's life after the tremors have ended. As these texts suggest within the realm of personal and collective experience, post-disaster futures start with everyday remaking; that is a process of making anew and weaving again of the ripped and complex fabric of one's life, knowing that many of its defining threads are not there and cannot be replaced.

In effect, the future that the narratives try to envisage from the various positions they take is an *unhealed* one. It is one which carries through and carries on the knotted work of healing as a multi-fold process. It encompasses, among other things, after Veena Das and Arthur Kleinman, 'engaging in repair of relationships in the deep processes of family, neighborhood, and community […] resuming the task of living (and not only surviving)',[13] knowing, at the same time, that 'while everyday life may be seen as the site of the ordinary, this ordinariness is itself recovered in the face of the most recalcitrant of tragedies: it is the site of many buried memories and experiences'.[14] Against illusions of an achievable wholeness, one that is often equated with the recovery of the pre-disaster past, this future-oriented perspective seeks to reconfigure and repair discourses and structures of vulnerability all the while recognising that even if some structures can be amended, the personal losses and absences will never fully be alleviated. Still, such lack of closure emphasises an open-ended outlook which allows and calls for a non-catastrophic conception of Haiti's future, one beyond equally confining notions of repetitive, cyclical or unalterable, linear and ongoing time of the disaster.

Narratives, and for Yanick Lahens novels in particular, give the reader 'sense of the world', creating a 'moving proximity' to and participation in one's life and live stories. These complement – are distinct from but not in opposition to – the types of knowledge and information about the world that hard and social sciences can provide. For the Haitian writer:

> When you look at social sciences, when you take statistics, economics, you have information about the world. But what will give you the sense, the taste, of the world? It will be the reading of a story of someone who lives biology and physics in his/her life. That we live too, and that we see, and that is why we feel moved when reading this proximity.[15]

The fictional and non-fictional accounts of the 2010 disaster, through their polyphonic form and sharing, to a varying extent, in their thematic preoccupations and ethical motivations, draw the reader into the experience of the earthquake. They hope to provide this 'taste of their world', that is the here and now as well as the aftermath and the reverberations of the January tremors.

Far from the sensationalist accounts of extreme suffering or exceptional survival, for the communities left behind, post-disaster present and future are characterised by the ongoing experience of structural violence, one that is countered by the everyday, unspectacular yet no less significant, ordinary practices of resistance and being in the world. For those who survived, as they attempt to find and live a daily life, trying to accept the shocking memory and loss as permanent, the enduring material destruction of the city space continues to be a visible reminder of the haunting experience. Both the post-disaster present and future are defined by the labour of *slow healing* – an open-ended and non-teleological process.[16] The challenge and difficulty of living in the post-disaster everyday lies precisely in having to confront and to battle against this seemingly ordinary – yet no less violent – vulnerability which halts and disrupts any attempts at linear recovery. In effect, the post-disaster everyday emerges as a liminal and intermediate site. The disaster is not over yet and still haunts the present, and this post-disaster present remains a work-in-progress; an attempt to formulate and slowly move towards a non-spectacular, non-catastrophic, everyday future.

With news-reports filled with images of yet another disaster unfolding in the Caribbean (e.g. the 2017 Hurricane Irma), it is easy to give into apocalyptic visions of Haiti and the wider region as one, ongoing crisis. Yet, rather than just demonstrating the relevance of disaster studies, such images and reports should push us, the scholars in the field, to rethink and reformulate the coordinates and ambitions for our inquiry towards futures that are less disastrous and, since this is the core of the issue, more just. Part of the book's undertaking is to engage an interdisciplinary analysis of notions of disaster, ruination, reconstruction and recovery; that is to go beyond critique. If critique is often founded on disagreement, then my book calls for a move away from the pointing out of what the various perspectives on disasters – whether those coming from quantitative or qualitative sciences or humanities – 'fail to see' *via* an acknowledgement of 'what we *do* do' that can be complementary and enriching towards, more importantly, an imaging of 'what we *could* do'. My monograph expands discipline-bound approaches, challenges confining characterisations of the Caribbean as a continuous catastrophe and formulates a bold, future-oriented, rigorous and critical practice of *rasanblaj* (as gathering in Haitian Kreyòl) that tries to bring together and transform the distinctive fields and approaches not only for the sake of the *argument* but for the sake of the living. And the dead. This is not a purely theoretical field of inquiry.

However, many, too many died on that fateful day. Unrivalled in its scale and impact, the 2010 Haiti earthquake was, that year alone, only one of many earthquakes around the world. Neither was it the first earthquake to shake Haiti. Previous earthquakes, for example in 1701, 1751 and 1770,[17] similarly to the 2010 tremors, have been ascribed to the Enriquillo-Plantain Garden fault zone; in Matthew J. Smith's gloss, 'on this fault-line lies history'.[18] Yet with an annual global average of 15 earthquakes of magnitude Mw7.0 or greater[19] and widely accessible online technology which is capable of tracking earthquake activity on a real-time basis,[20] what seems to distinguish the recent tremors is the scale of the comparative devastation that they caused. For example, the 2010 Chile earthquake, which struck the country on 27 February, despite having a magnitude of 8.8, and estimated depth of 22.9 km,[21] killed only 723 people with a further 25 going missing. In September 2010 an earthquake of similar magnitude to the Haitian one, and depth of 12 km,[22] struck just outside Canterbury, New Zealand.

There were numerous injuries but no fatalities.[23] Media commentators,[24] and viewers' donations for that matter, often engage in and are based on a type of 'comparison of suffering' where the severity and importance of one disaster over another, as well as the attention we give to it, are determined by lives lost. Yet, all lives are irreplaceable regardless of where the disaster happens. At the same time, a comparative look at the dissimilar effects of one particular hazard demonstrates the disproportionate damage endured by Haiti in 2010, clearly pointing to the ways in which pre-existing vulnerability and the wider pre-earthquake factors can transform an earthquake into a disaster of such scale and impact.

The devastation was beyond words and seemed, at least initially, beyond comprehension and representation with the January tremors challenging the meaning of history itself (Part II). Still, global media coverage was quick to turn to and reiterate tired tropes of Haitian and Haitians' exceptionalism, progress-resistant culture,[25] and crippling and, at once, ennobling poverty as ways of framing and explaining the January tremors and the devastation they caused. Yet for those who were directly affected by the disaster, witnessed its immediate aftermath on the ground or via direct and indirect communication with family members and friends – among them the authors analysed in this book – what happened on 12 January and in the subsequent days could not be reduced to or described by any such reductive tropes and images. No one experience, explanation or memory of the earthquake was the same. Each telling and re-telling of the story and the chain of events of that fateful day, a form of polyvalent witnessing that continues to unfold – each *narrative* of 12 January 2010 – bears a distinct perspective of what happened 'at the moment of', during the months and years preceding and following it; that is the past, present and future of the January *Goudougoudou*.

Narrative, understood more broadly and as used in this book, can be defined as 'a telling of some true or fictitious event or connected sequence of events, recounted by a narrator to a narratee'.[26] A narrative consists of 'a set of events (the story) recounted in a process of narration (or discourse), in which the events are selected and arranged in a particular order (the plot)'.[27] Each of these stories of the experience of the disaster, whether recorded,[28] written down, shared privately in interviews or publicly with potential readers, carries a distinct emotional power. Each, in a singular way and to varying ends, seeks to confront and control the experience of the disaster, and is a form of positioning in the world, calling forth an ethical-affective response. Each text calls for and provides novel ways of framing the 2010 disaster with its specific past, present and future. Listening to and reading the different voices and distinctive perspectives on the earthquake, it becomes clear that for those directly affected by the disaster, what happened on 12 January 2010 was a singular event. In its original, etymological sense from the *Oxford English Dictionary*, the earthquake was a catastrophe, 'an event producing a subversion of the order of things'[29] and 'a sudden turn, conclusion'.[30] Unprecedented in their scale, yet highly differentiated in their impact, the January tremors forever changed subjective and collective perceptions of past, present and future – now always interpreted in the light of the disaster.

With all its philosophical weight, for those affected this was '*bagay la*' ('the thing' in Haitian Kreyòl) as 'what had happened had no name and was so outside what we considered liveable or bearable, that it could not be named'.[31]

However traumatising it might be on a personal level, the January disaster was not a disconnected happenstance escaping all comparison and comprehension. Embedded within a web of relations the earthquake, at once, underscored and revealed previous societal, ecological and political fault lines.[32] In other words, the impact of the tremors was 'less the result of geophysical extremes' and more a '*function of ongoing social orders*, human-environment relations, and historical structural processes'.[33] Seen in more political terms, for Mark Schuller and Pablo Morales, the January disaster was a '*kriz konjonkti* – a conjunctural crisis',[34] that is 'the intersection of neoliberalism and foreign control, together with the complicity of Haiti's elite and government'.[35] At the same time, the word crisis as '*kriz*', in Kreyòl, and as '*la crise*', in French, also captures the sense of sudden interruption, not just an ongoing condition as the English usage seems to imply. This widened definition encompasses then, after Greg Beckett, 'an embodied condition that is rooted in the social and psychological experience of an individual'.[36] *Kriz*, also used to describe a moment of possession by the Vodou *lwa* (spirit), bears similarity to *sezisman* (shock) and is 'an emotional reaction to trauma'.[37] Both 'are unmediated bodily responses to loss'.[38] These ruptures of *kriz* 'are more commonly thought of as conditions that leave one vulnerable to illness or death by other means ... and are forms of embodied suffering'[39] that are the result of normalisation of vulnerability. The prolonged geopolitical context of structural vulnerability, the protracted past of the event that lives on and is lived through in the form of everyday depravations, helped transform the Haitian earthquake into a disaster on such a scale.[40] Its present is the immediate aftermath of 12 January; the future of the disaster is shaped by the material, socio-economic and cultural reconstruction processes that have ensued. There is no disaster without compound vulnerabilities,[41] with traceable and oft unnatural processes, such as economic and political exploitation, directly determining whether a natural hazard will become a disaster. This focus on constructed and sustained vulnerabilities is key for our understanding of an age of 'general disaster'[42] and 'the before and after of the [January 2010] catastrophe'[43] in Haiti.

Heterogeneous, contradictory and fragmentary as they are, narratives of the 2010 earthquake are a source of layered knowledge of the disaster – of the fateful day itself as well as of its embeddedness in long-term vulnerabilities – testifying to the ongoing, yet extremely differentiated, collective and personal impact of the tremors. Most of these individual stories and impressions of the earthquake were told to just the closest friends: some were collected and archived as oral histories of the disaster;[44] whereas others were shared with the wider global community as published narratives, among them the seven texts explored in this book. While they share an ambition to preserve and construct individual and collective memories of the event, the selected texts have distinct aesthetic, ethical and political aims and allow for an exploration of the multiple meanings of the earthquake, accounting for the histories which underlie the disaster as well as those which shape other literary engagements with it. In their immediacy and ability to capture the multifarious nature of disasters, these varied texts testify to the urgent need to give form and sense to an overwhelming experience, depicting, at the same time, realities that are otherwise omitted. In short, these narratives counter and diversify the silencing discourses of exclusively technocratic solutions to disasters. The book is attuned to the different contexts and traditions – literary, cultural, ideological – that inform each of the texts and positions the exploration of these fields within an interrogation of the extent to which narratives can account for the experience of this particular event, and of disasters more generally. A detailed analysis of these backgrounds to the disaster allows for a greater and more nuanced reading of selected texts. For example, it brings to the fore some of the ways in which these narratives intervene, or are implicated, in the broader context of humanitarian neoliberalism that has shaped international responses to the disaster and post-earthquake relief and recovery efforts. As such, these fictional and non-fictional texts, which greatly differ in terms of thematic concerns and formal complexity, complicate our sense of the event and offer a productive entry point into the exploration of these wider questions of how disasters are defined and how post-disaster reconstruction and recovery are practised in Haiti.

Instead of attempting a comprehensive analysis of all the works published since 12 January 2010 – a well-nigh impossible task in the context of the outpouring of literary responses to the tremors[45] – this book assembles texts characterised by their networked, multi-sited and polyvalent positioning, identifies among them key interpretative frames from which the 2010 earthquake has been approached, and emphasises the event's global translations, connections and repercussions to which each of the narratives testifies. The January tremors, its histories and the permanent imprint it leaves behind extend well beyond the boundaries of the nation-state; the 2010 disaster is not just a distant problem of a far-away nation, a separate event, *à part*, disconnected from everything else, but rather it is a full, and violent, manifestation, in its own right, *à part entière*, of ruinous, long-term, vulnerabilities. As such, the very character of the disaster calls for such a layered and polyphonic approach, one which reflects the processual, multi-scalar and uneven character of the events and their aftermath. These seven texts approach the disaster from radically different positions, cognitive spaces, theoretical frames of reference and meaning-making practices, including those of Haitian and non-Haitian survivors of the disaster, direct witnesses to the event, and those for whom writing was a way of countering the physical remove from the scene of the disaster and of creating emotional proximity to the victims of the earthquake. In other words, the choice of texts reflects the knotted nature of the disaster experience. In one blow, it thrusts and brings together unlikely voices and witnesses who, at once regardless and due to their previous positioning, are now at the centre of the disaster and whose lives centre around it. Still, the narratives share in their focus on the past, present and future of the January quake and their attempt to try and make sense of what happened and to give expression to the lived experience of the disaster, with some of them seeking ways to speak against 'discourses of vulnerability'[46] that often silence those most affected, rendering them voiceless. In effect, this assemblage of narrativised attempts to find a form and ways of responding to the January tremors is also indicative, in many ways, of the oftentimes discordant, if not cacophonous, efforts on the ground, and the overwhelming reality of the aftermath when one of the most difficult tasks is often to find out, something, anything of what had happened.

Uncertainty, confusion and the urge to account for and respond to – in whatever way and any ways that seem possible – the scale of devastation and loss cuts across the otherwise radically different positions and starting points for these narratives. These voices brought together here come from writings by well-established authors such as Rodney Saint-Éloi and Dany Laferrière, both from Haiti and living and working in Canada (Parts II and III), or Lionel-Édouard Martin, a prolific French diasporic poet and writer who lived at the time in Martinique (Part II), as well as those of first-time writers such as Sandra Marquez Stathis (Part III), an American former political observer who had worked in Haiti in the 1990s, and Dan Woolley (Part IV), a charity organisation worker who came first time to Haiti in the days before the earthquake, or Laura Rose Wagner, an anthropologist who lived and worked in Haiti from 2009 to 2012, and Nick Lake (Part IV), who followed the events as they unfolded from the UK, both for whom the genre of young adult novel offered the best form to respond to the earthquake and imagine Haiti's future otherwise. Neither makes claims to an exhaustive or complete knowledge of the disaster and its histories; no one experience is the same. In response, only such a comparative and immersive layering of their insights, as this book attempts to do, effectively portrays and can engage with the real dissonance of issues, interests, worries and personal and collective stakes and sufferings lived in the days, months and years after the initial tremors. These narratives cross the same city sites and often reference the same events yet occupy contrasting interpretative standpoints, revealing the dual personal-global impact, the many translations and the complex *connection*, rather than separation, between the 2010 earthquake, personal and collective ways of being embedded in and linked to the earthquake, of making sense of it, and the wider hemispheric and international politics. This book, in itself a site of encounter, occasions a meeting of these distinct, layered yet networked perspectives on the disaster experience that are otherwise kept separate, categorised under limiting labels of inside/outsider responses with many being dismissed, under a generalising label, as just 'do-gooders'' self-laudatory tales.[47] The earthquake, with one sudden blow – *tout d'un coup* – uproots and brings together these distinct voices, perspectives and ways of knowing, now all in the midst of the disaster experience.

The narratives I have chosen to read together try to grapple with a highly particular event, determined by multiple pre-existing factors, while providing an insight into the deeply personal effect the earthquake had on individual lives. In short, they cut across the scales and join the individual threads and fibres that make up the disaster fabric. From their immediate reflection on what just happened (Part II), through to the attempt to take in these seconds' devastating material impact (Part III) and the permanent imprint it has left on individual and community lives (Part IV). Analysed together, these highly diverse personal stories and imaginative responses to the disaster also give expression and form to the urgency to witness, to create circumstances so that an accretive and compound perspective on the event and its aftermath may take shape. They then point to the contrasting meanings and explanatory frameworks that are being called upon. Whereas some of the selected accounts emphasise primarily, in Toni Pressley-Sanon's gloss, 'vicarious traumatization (an emphatic reaction to witnessing)',[48] others point how witnessing to the disastrous event, as E. Ann Kaplan argues in other contexts, can lead to 'a broader understanding of the meaning of what has been done to the victims, of the politics of trauma being possible'.[49] In the process of doing so, each narrative reveals its limits and limitations in accounting for the scale of the devastation and loss. As they attempt to capture, give insight or imagine the lived experience of the disaster and the preceding reality of compound vulnerability, they equally give voice to the ongoing hurt and the unending labour of recovery, of remaking a life among and beyond the ruins of the city, day in and day out.

The texts' polyvalent efforts to come to terms with what has happened give a certain sense or order to the overwhelming experience, imagining Haiti's non-catastrophic futures as central to scholarly analyses of the 2010 tremors, which also seek to disentangle 'the knot' that is the experience of the earthquake. The image has a double significance. In the broader conceptual uses, a 'knot', as Roger Luckhurst defines it, is a heterogeneous concept whose use and permeation in a number of disciplines and historical and cultural contexts 'must be understood by the impressive range of elements that it ties together and which allows it to travel to such diverse places in the network of knowledge'.[50] Here, disaster is such a composite concept that braids together, among many other threads, embodied experience, structural, macro scales analyses, geological characteristics of an environmental phenomenon, all under one denotation. As Terry Rey and Karen Richman demonstrate, this sense of tying and bringing together is also central to the Haitian Kreyòl metaphor of *pwen*, 'based upon the French word *point*'.[51] Difficult to grasp due to 'its resonance with pre-modern perception',[52] *pwen*, after Karen McCarthy Brown, can be defined as 'anything that captures the essence or pith of a complex situation'[53] and 'reformulates it so that it can be easily grasped and remembered'.[54] *Pwen* is equally the ' "point" where two or more lines of power intersect [,] exists only as a relation',[55] is a site through which Vodou communication works[56] and 'can perhaps best be understood as a mode of embodied experience'.[57] Across their diverse uses, 'the knot' and *pwen* evoke points of contact between different epistemologies and ways of making sense of the world, with such a composite frame guiding this book's consideration of narratives of the 2010 earthquake.

In effect, the narrative form, sharing in and complicating the gesture of disentangling, does not just act as a representation of the experience of the disaster, whether lived, fictionalised or imagined. Rather, these texts' narrative form provides a distinct interpretative approach and starting point that hopes to help their readers grasp the complex nature of disasters and their aftermath. In other words, this book's turn to literary comparison and its close-reading enable a composite analysis, one which encompasses the social, political and individual dimensions of the disaster. The complexity of the analysis is heavily informed by the intricate and imbricated nature of a complex array of vulnerabilities which preceded the 2010 tremors and continue to shape the post-disaster everyday. Jointly, these threads, are organised around categories of time, space and self. These are foundational components of personal experience and building blocks of narrative form and also make up the interwoven fabric of the wider disaster experience: where, although in counterpoint, the distinctive voices converge in the violent polyphony of *douz janvye*. The three categories enable thus a cross-disciplinary treatment of narrative responses to the earthquake as literary texts *and* as interventions into the discursive and imaginary constructs of Haiti. This book, then, offers the kind of interdisciplinary methodology that might account best for the histories that underpinned the 2010 Haitian earthquake, and that might go some way towards explaining the narrative responses it inspired. In so doing, *Disasters, Vulnerability, and Narratives: Writing Haiti's Futures* directly answers the call made to Haitian Studies scholars to contribute to a brave new imagining of an empowered 'future for/with Haitians, one that can also be shared with the rest of the world'.[58] Literature and literary analysis can directly support this process by confronting history in new ways and gesturing to new histories, both of recent disaster and of the country's longer past.

'Not every windstorm, earth-tremor, or rush of water is a catastrophe [...] It is the collapse of the cultural protections that constitute the disaster proper'[59] with pre-existing factors and structures, as well as community's ability to respond, directly determining the development of a disaster. In other words, vulnerabilities not environmental phenomena and hazards, such as storms or tremors, are root causes of disasters.[60] Within this longer and integrated view, one that joins environmental, political history to personal experience and oft contrasting interpretative and sense-making frames, the pre-disaster past emerges and is best conceived as *multi-scalar vulnerability*. It is one that exists, manifests itself and cuts across multiple, non-exclusive levels and timescales. The 2010 Haitian earthquake is only one example of the ways in which this multi-scalar vulnerability, an aggregate of social, political and environmental fissures, amplifies the force of geological tremors, fatally transforming the quake into a catastrophe. The concept also captures the coming together of compound factors, embedded within and equally contributing to the overlapping dynamics of regional, hemispheric and global politics and histories.

In other words, Haiti's vulnerability is not a confined nor a confineable phenomenon, 'Haiti's own problem'. Rather, as a combination of long-term processes of a global character, environmental degradation and marginalisation, this multi-scalar vulnerability is a manifestation of uneven connectedness and a direct expression of a wider, not always immediately visible and potentially catastrophic, susceptibility to environmental hazards and phenomena, one that stretches well beyond Haiti's borders. Considerations of what happened, what followed afterwards and what led up to it, all allow for a detailed engagement with the knotted histories of the January event, and with the multiple factors which, combined, created 'the conditions for a disaster'.[61] Together, they signal the human-caused character of disasters,[62] testify to their accretive nature[63] and processual character[64] as well as their embeddedness in, rather than a radical separation from, the 'everyday normal' or a break from 'the rest of material life'.[65]

Disasters are not isolated and single events. Rather, they are slow-onset,[66] extended and complex crises.[67] In other words, they are 'the extreme situation which is implicit in the everyday condition of the population'[68] that requires 'creat[ing] ways of analyzing the vulnerability implicit in daily life',[69] It is precisely the everyday 'normal', defined by long-term vulnerability processes,[70] that increases risks and creates disasters, determining the scale of their impact[71] on individuals and communities. Personal accounts and fictionalised narratives of the 2010 disaster provide direct insight into such quotidian experience of differentiated vulnerability and its disastrous manifestations, with some texts boldly imagining how to act upon it. In short, the January disaster might have flattened the city but it did not level social inequalities and compound vulnerabilities which directly contributed to the scale of this processual disaster and its unfolding aftermath.

At the same time, while pointing to long-term and compound processes that create such complex unevenness and contribute to a population's marginalisation, my analysis is equally attuned to and critical of the ways in which 'the discourse of vulnerability' has functioned as a form of 'Western [d]iscourse',[72] classifying certain regions 'no less than the previous concepts of tropicality or development [...] as more dangerous than others'.[73] In effect, such discourses continue to divide the world 'into two, between a zone where disasters occur regularly and one where they occur infrequently[74] reiterating, under the guise of 'development', earlier colonial tropes of danger, marginality and backwardness used to describe 'non-Western' spaces, such as Haiti. Such reductive allocations of vulnerability fall short of accounting for the uneven entanglement of 'the north' and 'the south', the multifarious and differentiated exposure to and the resulting experience of risks, hazards and vulnerability, even within the same community.

Narratives of the 2010 earthquake make clear and testify to this uneven reality of stratification: life-saving, for some, and fatal, for most. Many of these texts, as forms of creative response to, in John McLeod's gloss, 'the deprivations of disaster', 'insist upon the demands of justice as part of their critical activism and their commitment to community, ecology and democratized recovery amidst the slow violence and heavy modernity of our post-progressive present'.[75] In other words, these accounts attend to the lived realities of overlapping 'incremental and accretive'[76] violences that, as Rob Nixon emphasises, are played out, experienced and belatedly manifested 'across a range of temporal scales':[77] from increasing food insecurity and lack of adequate housing to the sense of seemingly ever-present threat to livelihoods, whether due to rising sea levels, seasonal storms or environmental hazards that become disastrous. In sum, these fictional and non-fiction accounts allow us to go beyond a critique of single disciplinary frames or respective emphases towards a formulation of an interdisciplinary, *hinged* approach to the 2010 disaster, one that encompasses personal experience of it within a rigorous scrutiny of its longer pasts and unfolding present and future.

Hinged chronologies

The exploration of the multiple histories of the 2010 earthquake and the many roots of the January disaster, as well as their quotidian and abrupt manifestations to which many of the narratives of the 2010 earthquake testify, points to, what I call, '*hinged chronologies*' of the disaster; that is, the compound processes that contributed to the disaster as well as varying points of connection between this and earlier historical events. At the same time, the term captures the varying explanatory frames of reference, e.g. spiritual traditions or personal and collective history, that are mobilised in an attempt to give meaning and sense to an overwhelming experience and its haunting reverberations. In his narrative of the earthquake, for example, Rodney Saint-Éloi (Part II) draws a clear connection between the 2010 event and what he sees as the first tremors in the island's history, namely Columbus's arrival in 1492. For the Haitian diasporic writer who survived the January disaster, the aftershocks of the first historical quake only amplified the impact of the geological tremors in 2010. Speaking from a very different perspective, one rooted in anthropological approaches to disasters, Anthony Oliver-Smith conceptualises the 2010 earthquake in strikingly similar terms. For him, what Haiti experienced on 12 January was in some respects, 'the culmination of its own more than 500-year earthquake'.[78] Whereas Rodney Saint-Éloi's comparison complements in many ways Kamau Brathwaite's earlier formulation of the Middle Passage as 'the original explosion' and an 'ongoing catastrophe',[79] Oliver-Smith's metaphorical statement, for its part, serves to illustrate his argument that disasters 'are processes that unfold through time, and their causes are deeply embedded in societal history'.[80]

For both Oliver-Smith and Saint-Éloi, however, the reference they make to colonialism and Haiti's postcolonial condition is more than just an attention-catching figure of speech. As used here and throughout the monograph, the term 'postcolonial' refers to the wider theoretical approach rooted in humanities-based critical attention to the long-term legacies of colonialism and imperialism and their constructs, not just a defined historical moment. Haiti is the world's first black republic and gained its independence from France in 1804, becoming an independent, sovereign and non-colonial state. Yet, as taken up for example in humanities-based critical school of postcolonial studies,[81] the term 'postcolonial' does not just refer to Haiti's post-1804 history. Rather, the concept emphasises the longer legacies of colonialism, imperialism, often visible in other forms, and is concerned with the broader questions of power, hegemony, economics, culture and ecology as played out, for example but not exclusively in, regions defined by colonial pasts.

The term, whether taken up implicitly or explicitly in these responses to the 2010 earthquake as well as in the overarching task of thinking through the histories of disaster, 'is as much about conditions under imperialism and colonialism proper as about conditions coming after the historical end of colonialism'.[82] Rather than denoting some fixed and chronologically suppressed moment, post-colonialism is, then, better conceived as a *process of postcolonializing*[83] and an anticipatory discourse, a future-oriented critical approach that seeks to 'relate modern-day phenomena to their explicit, implicit or even potential relations to this fraught heritage'[84] of the colonial aftermath. Speaking from different disciplinary standpoints as well as varying levels of personal attachment to Haiti, Oliver-Smith, Brathwaite and Saint-Éloi all recognise and reiterate the importance of these after-effects of the colonial experience for the understanding of disasters in the region. In so doing, all three call for an epistemic shift in thinking through the past, present and future of disasters; that is a composite approach which acknowledges the indelible marks of colonialism in the region, and is attentive to long-term processes of vulnerability construction and structural violence which defined the pre-disaster everyday and continue to constrain Haiti's post-earthquake reconstruction.

Building on these radical rethinkings of disaster as well as Gina Athena Ulysse's insistent call that 'Haiti needs new narratives',[85] I craft such a compound, interdisciplinary and longitudinal approach to the 2010 earthquake and its aftermath. Such a multi-frame inquiry builds upon vulnerability-centred approaches to disasters as well as Haiti-specific analyses of vulnerability production, with narratives of the 2010 disaster providing key insights into the realm of the personal and lived experience of vulnerability. This allows, in turn, for a novel reformulation of the notion of disaster time and space as well as visions of post-disaster ruination, reconstruction and recovery. While focusing around these concepts, the chapters that follow also directly confront and challenge disempowering and recurring discourses on Haiti. In binary terms, these combine the impermanence of the failed Haitian state, 'unable to properly govern itself, done in by itself and acts of nature',[86] with the permanence of its ruination, placing them alongside visions of the country and its people as, at once, voiceless in their poverty and too vocal in their demands for a fair and just today and tomorrow.

Moreover, in contrast to eternalising visions of 'an irremediable disaster',[87] *Disasters, Vulnerability, and Narratives: Writing Haiti's Futures* challenges such framings which simplistically equate continuity – for example, of practices of exploitation and marginalisation or processes of ecological degradation – with unavoidability. Such extreme positioning of contemporary, post-earthquake Haiti, seen by Martin Munro, as 'but a presage of a broader catastrophic collapse, a window onto all of our apocalyptic futures',[88] is ultimately a depoliticising one. In this rereading, the recent disaster and its aftermath are just another example of the country's many failings, easily fitting into the compound narrative of repeated catastrophes that, in one version or another, has long been attached to Haiti. Within such apocalyptic imaginaries, crisis in Haiti, as Greg Beckett argues, 'is not just a regular feature of Haitian society but a *normative* one, an inherent, inescapable condition that plagues the country and its people' [italics added].[89] Or, in Kaiama L. Glover's words, Haiti is presented as 'more or less all earthquake, all the time'.[90] While giving voice to the sense of urgency and the need to look to long-term histories of disaster, the paradoxical evocation of apocalyptic solidarity seems to ignore and erase the reality of uneven connectedness, compound vulnerabilities and asymmetrical exposure which define the global present and shape possible futures for the different regions. In effect, this confining vision of an unstoppable and joint descent into some apocalyptic future risks to further normalise 'Haiti within paradigms of disaster that reproduces vulnerability and marginalization on the national and international stage'.[91] Finally, such a rhetorical emphasis on an unstoppable collapse detracts from questions of justice, agency and responsibility, undermining the importance of the struggle towards *disaster-free futures*; that is a non-idealised time of thinking and understanding Haiti otherwise than as an exception – excess or abject – or as a disaster space, always in the extreme.

The commitment to forging and working towards Haiti's non-disastrous futures is central to this book, driving its inquiry into the disaster's long histories and its violent present manifestations, seen in non-fatalistic terms. In this context, this book's detailed engagement with narratives of the 2010 earthquake is a starting point for its contribution to the linguistic and conceptual transformation and expansion of disaster studies, one that, seconding Anthony Carrigan's call, would shift the field away from its 'complicities with militarism, neocolonialism, and capitalist exploitation, and towards the emancipatory vision of recovery and "healing"'.[92] Such emancipatory recovery challenges reversely teleological perspectives. In the apocalyptic mode, they suggest Haiti's unalterable decline, viewing the country's present and its past as 'a temporality defined by rupture and breakdown'.[93] In their prescriptive didactic and ideological inversions, they paint a clear vision of what Haiti *should* be like. Both, however, only reinforce, according to Gina Athena Ulysse, Haiti's entrapment in 'singular narratives and clichés that, unsurprisingly, hardly moved beyond stereotypes';[94] Haiti 'in the public domain was a rhetorically and symbolically incarcerated one'[95] and often continues to be so. Contrary to such prescriptive and punitive visions, an ongoing, work-in-progress, *recovering* is, in the present continuous mode, an open-ended, hinged and multi-scalar process, unfolding at multiple, non-synchronous speeds.

Voicing catastrophe

Those directly affected by the disaster as well as those following it from afar might be hurt but they are not voiceless. Narratives of the 2010 tremors give expression to and articulate the deeply heterogeneous lived experience of this extended temporality of the disaster and its longer histories. In the process, these narratives try to give meaning to memories of the past, experience of the present and ideas of the future, both personal and collective, that are all connected to and enmeshed within wider processes of vulnerability production and its disastrous manifestations. Emphasising the importance of literary production to a deeper understanding of the experience of disaster, for Mark D. Anderson, in his *Disaster Writing: The Cultural Politics of Catastrophe in Latin America* (2011),[96] post-disaster literature is not only a useful comparison but is itself a productive source of knowledge of the disaster and the politics behind different cultural representations of it. The so-called 'natural' disasters, in his view, 'give rise to powerful cultural and political discourses',[97] literary production among them, which provide competing explanations of each particular event. Narrativisations of disaster partake in this process of negotiation of meaning and authority and of creating an open-ended, unfinished archive of the earthquake with the fragmentary aesthetics of some of the works shattering, as Rachel Douglas's analysis of post-earthquake Haitian writing suggests, 'any notion that "all" of the story of the earthquake can be told within the bounds of a single artistic work or book'.[98] By pointing to the multiple histories which underlie a specific disaster, they can disclose and challenge existing power dynamics, offering new ways of remembering the event.[99]

Moreover, as Anderson argues, literary responses usefully indicate the collective importance of a given disaster and its political and cultural significance:

> The volume of literature written about a particular disaster and the involvement of canonical literary figures are prime indicators of its [disaster's] cultural and political relevancy for the nation. Likewise, more developed and sophisticated cultural production on a disaster allows for a deeper reading of the forces at work. It is for this reason that I do not study the recent (2010) disasters in Haiti and Chile. As horrifyingly mesmerizing as they may be, a modicum of historical distance is necessary to judge the implications of literary and artistic production for political change.[100]

In this extended quotation, Anderson establishes a very clear relationship between specific kinds of literary production and their value as a source of knowledge. However, the binary differentiation between 'high' and 'low' literature, marked here, results in an overlooking of 'less sophisticated' forms of expression, which also enrich one's understanding of the disaster and its aftermath by giving voice to oft excluded or dismissed experiences and perspective and, as such, should not be dismissed so easily.

In addition, in direct reference to the 2010 Haitian earthquake, Anderson problematises the timely character of many of these post-seismic interventions and insists on the importance of 'a modicum of historical distance' – a temporal remove from the event. Certainly, the apprehension of disasters changes throughout time, but 'historical distance' is not in itself a guarantee of a work's quality and sophistication or that of the insights it can offer. Here and elsewhere within Anderson's framework the emphasis is placed on the function of literary texts rather than their generic and aesthetic qualities. As a result, the value of literature is seen primarily in terms of its potential to shape the cultural politics of disaster, namely its influence on ways of remembering the event and the discourses surrounding it, and its ability to expose the 'deeper forces at work' that define a particular event.

Going beyond this limited focus on the potential political function and uses of post-disaster literary production, this book sees narratives of the 2010 earthquake as challenging ideas of disaster as a space and time-confined event as well as problematising teleological and closed notions of reconstruction and recovery that hope for a, seemingly possible, attainable and desirable return to the pre-disaster state as their goal. Rather, the texts analysed here reposition the January 2010 jolts as, at once, a deeply personal – unprecedented in its emotional weight – event and a coming together and violent manifestation of long-term processes and vulnerabilities. In effect, these texts call for a layered, non-binary approach to the 2010 disaster, one that is rooted in the specificity of the event, 'native *arguments*'[101] [italics original], and encompasses the discordant and multifarious experiences of the disaster, without, however, reducing them under homogenising labels – 'local' or 'native' voices versus 'outsider' perspectives – or seeing them, in oppositional terms, as simply confirming or challenging 'cultural and political discourses' on disasters.[102]

Finally, disaster, its past, present and future, is a profoundly human experience and a lived reality which cannot be captured under such binary formulations. As sudden and tangible manifestations of long-term vulnerabilities, disasters are also defined by the scale of their impact on given communities, leaving a permanent imprint on individual lives; in other words 'without people there can be no disaster'.[103] Or, as Rodney Saint-Éloi (Part II) puts it in his memoir: 'in a general disaster, there is always a disaster of the self'.[104] It is this human impact, often captured in reductive terms as a sum of fatalities and monetary losses, that further differentiates environmental and geological phenomena, such as storms or the movement of fault lines, from disasters they can become. In sum, as expressions of living in times of disaster and as forms of *arguments*, of analytical and not just illustrative value, narratives of the 2010 Haiti earthquake interlace the long pasts and violent present of disaster and try to envisage non-catastrophic futures, taking one's enmeshed experience – as lived through, witnessed or imagined – as a starting point.

Writing in the aftermath

The process of narrativising the January tremors then becomes a search for a form and language which can approximate this testing experience and communicate the particularity of these at once local and global, hinged Caribbean histories of disaster. Kamau Brathwaite famously claimed that '*hurricane does not roar in pentameter*'.[105] The Barbadian poet insisted on the need to find a new form of literary representation and expression, rooted in histories of imperialism, that might reflect the ecologies and histories of the Caribbean region. Similarly, post-earthquake narratives, to a varying extent and with different results, attempt to create a narrative form which reveals and represents the experience of the disaster and the impact it had on individuals and communities, locally and globally. While providing testimony for traumatic experience and an insight into narrative politics, disaster accounts of the kind examined here try to grapple with the composite nature of disasters and expose the many contrasting meanings of the January tremors that cannot be captured by methodologically confined approaches. These texts, in their immediacy and ability to capture the multifarious nature of disasters, can in many ways anticipate socio-scientific analyses, depicting realities that are otherwise omitted while allowing the readers to explore the global dimensions of the earthquake and its many narrative translations both within and outside of Haiti.

The immediate aftermath of the earthquake and the months following upon it saw an outpouring of literary creativity, as enumerated above, in the form of memoirs, novels, collections of short stories or anthologies. Easily accessible self-publishing opportunities as well as the presence of many literary figures for the *Étonnants Voyageurs* literary festival, which was meant to take place that same week in Port-au-Prince, contributed to the proliferation of highly diverse post-seismic literary responses. Formal and conceptual heterogeneity characterises these narrative takes on the disaster. Among them are survivors' accounts, written both by Haitians and non-Haitians (aid and charity workers, members of the rescue forces), as well as narratives of those who visited Haiti in the immediate aftermath of the disaster, subject to the anthropologists Marc Schuller and Pablo Morales's strongest criticism. Varied as they are, these diverse fictional and non-fictional representations reflect the importance of narrative writing as a means of coming to terms with global events while testifying to the profound sense of interconnectedness between subjective experiences of the event and their wider collective, political and historical significance.

This sense of interconnectedness and heterogeneity of experiences of and perspectives on the disaster is also emphasised in two early anthologies: *Haiti Rising: Haitian History, Culture and the Earthquake of 2010* (2010), edited by Martin Munro,[106] and *How to Write an Earthquake: Comment écrire et quoi écrire/Mo pou 12 janvye* (2011), edited by Pierre Beaudelaine and Nataša Ďurovičová.[107] Whereas the former brings together a series of short personal accounts and scholarly analyses of the earthquake, suggesting the importance of a joint view, the latter focuses primarily on post-earthquake responses written by Haitians living in Haiti or in the Diaspora. Beaudelaine and Ďurovičová's anthology explicitly underlines the responsibility of the author and the significance of literary production in the aftermath of the 2010 earthquake, while similarly pointing to the connection between the wider ethics of writing and the deeply personal commemorative function it has:

> Writing bridges the two times and provides a shelter for memory, individual and collective. Through poetry, fiction, and nonfiction, it traces the ethos of pain and care that affected everyone on the threshold of the fault line.[108]

Here, the poetic formulation points to writing's potential to offer an interpretative framework and ethical standpoint from which to approach the earthquake and to present and preserve the many memories and impressions of the disaster. This act of braiding together the different voices creates a polyphonic anthology while being, at the same time, a gesture of solidarity and an expression of the anthology's direct commitment to Haiti's reconstruction: 'The authors of *Mo pou 12 Janvye* keep the memory of Haiti alive, and beyond their individual differences, beyond all political affiliations, they have come together [...].'[109]

The various narratives assembled in *How to Write an Earthquake* thus see themselves as contributing to the country's recovery efforts and as fostering a spirit of unity and hope against everyday challenges and frustrations:

> Today, after the earthquake, the ongoing cholera epidemic, the indefinite duration of life in tent camps, the disappointments of the international community in spite of genuine global solidarity, and the general uncertainty about a democratically elected government, the narratives and the songs that tell the stories of the departed and the survivors are still very poignant and central, and remind us of the need for unity and forging forward together. Nou la! Nou ansanm![110]

Here, the aims of the anthology are made clear, with the collective impact of these contributions being considered more important than individual authors' ambitions for each text. Through its commitment to commemoration and remembrance, *How to Write an Earthquake* attempts to counter a prevailing sense of anxiety and disappointment without, however, negating the suffering and the loss endured by the Haitian people. Its call is clear: we are here, we are together. Sharing the concerns of some of the narratives discussed in this monograph, the two earlier published volumes hope to preserve the many distinct memories of the event and its impact, to demonstrate its complexity and to express their solidarity with and commitment to Haiti.

Even if cursory, this overview of the literary responses published so far reveals varying motivations and distinct aesthetic, ethical and political aims behind each narrative. While sharing some thematic preoccupations, literary responses to the 2010 earthquake 'insist on the freedom to express those themes in original ways'[111] that best correspond to individual authors' aims and aspirations. These texts I have brought together offer a means of pausing to reflect upon lived experience by forging a narrative space that brings together an awareness of the difficulty of accounting for exactly what happened and the impossibility of forgetting it.[112] This tension-ridden 'space of reflection', as Emmanuelle Anne Vanborre argues in a later collection *Haïti après le tremblement de terre: La forme, le rôle et le pouvoir de l'écriture* (2015), allows readers to 'reflect and think without pretending to offer the reality of an experience which we had not experienced'.[113] In addition, literary interventions, for Martin Munro, constitute 'one of the most striking and important means of communicating the effects of such a disaster',[114] being 'one of the most privileged ways for the outsider in particular to begin to comprehend the experience of living in and through a time of catastrophe'.[115] Yet whereas Munro sees *Haitian* literature as a privileged source of knowledge about the disaster and its histories, my analysis complicates the classic insider/outsider binary to argue for the value and importance of diasporic and 'foreign' narrativisations of the earthquake: these, too, shape the cultural frameworks through which such complex events as the January 2010 tremors are narrativised and apprehended.

Moreover, narrative representations of disaster, with varying levels of self-reflexivity determined partly by the generic conventions they use, draw on their potential to shape readers' affective investment and response, an issue particularly relevant to post-earthquake writing. To trigger or condition their readers, these narratives can take the form of a direct appeal to the reader, as employed in Woolley's conversion narrative (Part IV), or can be embedded within the narrative design and the relations of empathy it establishes within the text and between the text and the reader. Such accounts may also contribute to personal/collective emancipation and liberation by reflecting on both the personal and collective experiences of disaster through narrative form and demonstrating the limitations of fixed victim discourses and forging, in effect, non-hierarchical empathetic relations and by giving voice and agency to those directly affected by the disaster.

Overlapping in some of their aims, the narratives of the earthquake all demonstrate formal and conceptual limitations that relate to both narrative ambitions and textual aesthetics. These arise primarily from the difficulty of sharing the personal experience of the traumatic event and of engaging with its multiple contexts through narrative form without silencing or objectifying other victims' voices at the time when the disaster continues to unfold in different guises. First, it is important to note that while many of the texts have an autobiographical character, they are first and foremost literary *representations*; that is, they depict, mediate and make present a subjective experience, whether real or imagined, with the intention of comprehending the 'again' nature, the memory, the rewriting and reshaping of it. As such, they are necessarily partial and contestable. As attempts to write *témoignage*,[116] at once an account of and a testimonial to the disaster, they demonstrate the difficulty of reconciling the mix of 'irreconcilable' fictional and real elements that constitutes testimonial writing.[117] Narrative focalisation and voice, as reflected in some of the discursive tropes employed across the selected texts, pose an additional challenge for these composite narratives which, against the objectifying images of wounded black bodies and anonymous masses – presented, in Sibylle Fischer's words, as '[p]eople without names, without history, without location: mere bodies, all black'[118] – aspire to provide an empowering vision of post-earthquake Haiti.

Similarly, literary narratives, such as those analysed in this book, can interrogate the aforementioned discursive constructions of Haiti's wider history, culture and politics, showing how these were employed to explain the earthquake while never fully disentangling themselves from the circumstances and conditions under which they are reproduced. However critically intended, the texts discussed here often reveal their complicity with dominant discursive framings of Haiti, demonstrating insufficient critical assessment of their own form, tone and language, and sometimes falling into the re-enactment of an objectifying tourist gaze. For Andrew Leak, scholar of French and Francophone Studies, most post-earthquake narrativisations of the disaster risk enacting such a voyeuristic gaze which appropriates the traumatic experience of the earthquake. In his view, literary responses to the January tremors by a number of Haitian Francophone writers in particular, including Laferrière and Saint-Éloi, are complicit in the creation and affirmation of negative discursive constructions of Haiti as well as the economic and military recolonisation of Haiti over the last decade.[119] Although pointing usefully to the question of diasporic privilege, Leak's analysis seems to dismiss Laferrière's and Saint-Éloi's heightened awareness of the complexity of the disaster and its history and their own complicated commitment to Haiti. Equally, Leak makes little distinction between the manifestly different positions Evelyne Trouillot, Yanick Lahens, Dany Laferrière or Rodney Saint-Éloi write from, and the equally obvious formal differences between immediate online responses – like those necessarily emotive if overly hopeful *témoignages* in the French progressive *Libération* by Lahens and others[120] – and later published narratives. But the workings of these texts as *narratives* remain largely unexamined. After all, they are themselves reconstructions, instruments of control and vehicles for personal and cultural memory. In each narrative, the experience of disaster 'as it happened' is retrospectively revised and reinterpreted in the light of earlier, previously inaccessible knowledge and information. Consequently, the different narrative engagements discussed throughout the book all run the risk of reproducing reductive framings of Haiti, in general, and the 2010 earthquake in particular. Thus, they have the potential to contribute to further simplifications and mythologisations of Haiti's past and present predicament; they can create new myths or reinforce old ones.

Furthermore, in seeking to disentangle 'the knot' that is the experience of the earthquake, these texts try to chart a progression of events and to provide a sense of formal and thematic resolution – an impossible task. In the case of post-earthquake narratives, narrative *dénouement*, where earlier enigmas are resolved and conflicts unravelled, is inherently an unreachable goal. The 2010 tremors cannot be bracketed off in this way. Neither the event nor Haiti itself can simply be 'solved' nor 'saved'; the disaster is still ongoing in the different forms taken by post-earthquake reconstruction. This extended temporality of the event, now manifest through its social, cultural and ecological repercussions rather than physical aftershocks, disrupts teleological narrative designs of this kind. In sum, the challenge, both formal and conceptual, for these narratives is to go beyond tired metaphors of Haiti's unique ruin and degradation, to expose their dominative function and to move towards *synecdochic* modes of representation of Haiti and the 2010 earthquake that position the country and its predicament as a part of a bigger whole and in terms of contiguity, proximity and connectedness between the event, its context and wider global history. One of the most important tasks that postcolonial disaster studies and Haitian Studies set themselves is confronting and questioning such persistent discourses on Haiti as an unalterable ruin and unending disaster – a challenge and a task this book also explicitly confronts and takes up.

Notes

1 'Qu'est-ce qu'une minute, celle qui fait le compte rond? Pas grand-chose dans la vie d'un homme. Mais il est, quelquefois, des minutes qui dilatent à l'extrême leurs secondes et leur confèrent une épaisseur de gravats et de mort.' Lionel-Édouard Martin, *Le Tremblement: Haïti, 12 janvier 2010* (Paris: Arléa, 2010), p. 120.

2 US Geological Survey, National Earthquake Information, 'Magnitude 7.0 Haiti' (US Geological Survey: National Earthquake Information Center, 2013) http://earth quake.usgs.gov/earthquakes/eqinthenews/2010/us2010rja6/#summary [accessed 02 April 2013].

3 The USGS estimates the epicentre of the magnitude-7.0 quake at a depth of 13 kilometres. USGS, 'M 7.0 – Haiti Region', *Earthquake.usgs.gov* https://earthquake.usgs.gov/earthquakes/eventpage/usp000h60h#scientific [accessed 17 October 2017].

4 Laura Zanotti, 'Cacophonies of Aid, Failed State Building and NGOs in Haiti: Setting the Stage for Disaster, Envisioning the Future', *Third World Quarterly*, 31 (2010), 755–71 (p. 756).

5 'Haiti Earthquake Facts and Figures', *Dec.org*, www.dec.org.uk/articles/haiti-earthquake-facts-and-figures [accessed 6 September 2017].

6 Center for Economic and Policy Research, *Haiti by the Numbers, Four Years Later* www.cepr.net/index.php/blogs/relief-and-reconstruction-watch/haiti-by-the-numbers-four-years-later [accessed 20 January 2014]. Guglielmo Schininà and others, 'Psychosocial Response to the Haiti Earthquake: The Experiences of International Organization for Migration', *Intervention*, 8 (2010), 158–64 (p. 58).

7 Patrick Sylvain is clear in his assessment of President René Préval's [President at the time of the earthquake] silence in the days following the disaster, likening it to the failure to contain violence:

> It is the executive politician at a time of national crisis who might bring about hope and restore a sense of collectivity in the nation through his effective and sensible language. He might efface the potentially vile rhetoric that would further soil the nation. Instead, Préval did not fulfill his role as the national rhetorician, and he failed to provide the people with the grammar of hope, the inspirational language they were looking for. As a result, he failed to contain violence.
>
> Patrick Sylvain, 'The Violence of Executive Silence', in *The Idea of Haiti: Rethinking Crisis and* Development, ed. by Millery Polyné (Minneapolis and London: University of Minnesota Press, 2013), pp. 87–111 (p. 91)

8 *Goudougoudou* is an onomatopoeic approximation of the sound of the earthquake and is now used to refer to the 12 January 2010 disaster. Nancy Dorsinville, 'Goudou Goudou', in *Haiti After the Earthquake*, ed. by Paul Farmer (New York, NY: Public Affairs, 2011), pp. 273–81.

9 Veena Das and Arthur Kleinman, 'Introduction', in *Remaking a World: Violence, Social Suffering and Recovery* (Berkeley: University of California Press, 2001), pp. 1–31 (p. 20).

10 Here, I draw on the rich symbolism of *kalfou*, as a crossroads in Vodou and the site were Legba resides (Master of the Crossroads), and, after Claudine Michel, as a methodology for Haitian Studies. She writes:

In Vodou, *kalfou* is the site where Legba, spirit of change and transformation, appears before humans. The *kalfou* presents another way of approaching research and interventions on the ground. It is simultaneously notions, methodologies, epistemologies, imaginaries and projects that intersect. It invokes interconnectedness and a multitude of choices; it inspires us to reflect on continuing paths as well as interruptions and changes in direction. I regard the *kalfou* as an ideal symbol for new methodological approaches to Haitian Studies. Based on a Haiti construct and a sound epistemological foundation of native (and *diasporic*) origin, the *kalfou* takes into consideration Haiti's positions as a crossroads, historically and since the quake.

Claudine Michel, '*Kalfou Danje*: Situating Haitian Studies and My Own Journey within It', in *The Haiti Exception: Anthropology and the Predicament of Narrative*, ed. by Alessandra Benedicty-Kokken, Kaiama L. Glover, Mark Schuller and Jhon Picard Byron (Liverpool: Liverpool University Press, 2016), pp. 193–208 (p. 206)

11 In the words of Isak Winkel Holm, disaster accounts of the kind examined here can also offer a 'testing ground'. Isak Winkel Holm, 'Earthquake in Haiti: Kleist and the Birth of Modern Disaster Discourse', *New German Critique*, 39 (2012), 49–66 (p. 50).

12 'bearing, n.' *Oxford English Dictionary Online* (2012), www.oed.com/view/Entry/16571 [accessed 15 February 2018].

13 Das and Kleinman, p. 4

14 Das and Kleinman, p. 4. It is also equally important to recognise 'this entanglement of the larger political environment in both the acts of violence and in the creation of possibilities for healing'. Das and Kleinman, p. 19.

15 Quand vous allez maintenant du côté des sciences sociales, si vous prenez les statistiques, l'économie vous avez des informations sur le monde. Mais qu'est-ce qu'il y a vous donnez la saveur du monde? C'est du lire l'histoire du quelqu'un qui vit et la biologie et la physique et les statistiques dans sa vie. Hors nous vivons aussi, donc c'est parce qu'on voit, qu'on est en train de lire cette proximité qu'on se sent touchés.

[Unless otherwise stated, translations are my own.] Yanick Lahens, interview with the artist, 22 July 2017.

16 Kasia Mika and Ilan Kelman, 'Shealing: towards open-ended recoveries?' [article in progress; forthcoming].

17 William H. Bakun, Claudia H. Flores, and Uri S. ten Brink, 'Significant Earthquakes on the Enriquillo Fault System, Hispaniola, 1500–2010: Implications for Seismic Hazard', *Bulletin of the Seismological Society of America*, 102 (2012), 18–30; Monique Terrier and others, 'Revision of the Geological Context of the Port-Au-Prince Metropolitan Area. Haiti: Implications for Slope Failures and Seismic Hazard Assessment', *Natural Hazards and Earth System Sciences*, 14 (2014), 2577–87.

18 Matthew J. Smith, 'A Tale of Two Tragedies: Remembering and Forgetting Kingston and Port-au-Prince', Shaking Up the World? Global Effects of Haitian Tremors: 1791, 2010 (Aarhus University, Denmark; 10–12 August 2017).

19 British Geological Survey, *Earthquakes* www.bgs.ac.uk/discoveringGeology/hazards/earthquakes/home.html [accessed 21 January 2014].

20 One such example is *Earthquake Track* http://earthquaketrack.com/ [accessed 21 January 2014].

21 Zacharie Duputel, Luis Rivera, Hiroo Kanamori and Gavin P. Hayes, 'W phase source inversion for moderate to large earthquakes', *Geophysical Journal International*, 189 (2012), 1125–47. Here, the uncertainty is estimated at ±9.2., see: USGS, 'M 8.8 – offshore Bio-Bio, Chile', *Earthquake.usgs.gov* https://earthquake.usgs.gov/earthquakes/eventpage/official20100227063411530_30#executive [accessed 18 October 2017].

The estimations of depth range from 44.8 km to 22.9 see 'M 8.8 – offshore Bio-Bio, Chile', *Earthquakes.usgs.gov*.

22 USGS, 'M 7.0 – South Island of New Zealand', *Earthquakes.usgs.gov* https://earth quake.usgs.gov/earthquakes/eventpage/usp000hk46#origin [accessed 18 October 2017].

23 Here, the disparity between the scale of the loss and the ensuing devastation could not be more striking and instructive. Even though tremors 500 times more powerful shook Chile, Chile's earthquake claimed 723 lives. 'True, Chile's earthquake was farther away from a major city. An earthquake of the same magnitude as Haiti's struck just outside the major city of Canterbury, New Zealand, in September 2010. No deaths were recorded.' Mark Schuller and Pablo Morales, 'Haiti's Vulnerability to Disasters', in *Tectonic Shifts: Haiti Since the Earthquake*, ed. by Mark Schuller and Pablo Morales (Sterling, VA: Stylus Publishing, 2012), pp. 11–13 (p. 12).

24 For some examples of this highly problematic politic of comparison, with the non-survivors of the two disasters 'competing' for media space and attention, see, for example: Anthony Reuben, 'Why did fewer die in Chile's earthquake than in Haiti's?', *BBC News*, 1 March 2010, http://news.bbc.co.uk/2/hi/americas/8543324. stm [accessed 13 February 2018]; Tim Padgett, 'Chile and Haiti: A Tale of Two Earthquakes', *Time*, 1 March 2010, http://content.time.com/time/world/article/ 0,8599,1968576,00.html [accessed 13 February 2018].

25 One example of an objectifying approach, which uncritically mirrors colonial and neoliberal discourses, is David Brook's article 'The Underlying Tragedy'. In it, Brooks emphasises inefficacy of international aid and its lack of real impact, asserting that

> Haiti, like most of the world's poorest nations, suffers from a complex web of progress-resistant cultural influences. There is the influence of the voodoo religion which spreads the message that life is capricious and planning futile. There are high levels of social mistrust. Responsibility is often not internalized. Child-rearing practices often involve neglect in the early years and harsh retribution when kids hit 9 or 10.
>
> David Brooks, 'The Underlying Tragedy', *New York Times*, 15 January 2010, www.nytimes.com/2010/01/15/opinion/15brooks.html?_r=0 [accessed 25 February 2014]

26 Chris Baldick, 'narrative', in *The Oxford Dictionary of Literary Terms Online* (Oxford University Press, 2008) www.oxfordreference.com.wam.leeds.ac.uk/view. 10.1093/acref/9780199208272.001.0001/acref-9780199208272-e-760 [accessed 07 January 2014].

27 Baldick, in *The Oxford Dictionary of Literary Terms Online*.

28 *Haiti Memory Project* http://haitimemoryproject.org/ [accessed 14 May 2015].

29 'Catastrophe', in *Oxford English Dictionary Online* (2012) www.oed.com/view/Entry/ 28794?redirectedFrom=catastrophe#eid [accessed 12 December 2012].

30 'catastrophe', in *Oxford English Dictionary Online*.

31 The use of the term here is a direct reference to Nancy Dorsinville's analysis of the scale of the disaster and its collective impact. '*Bagay la*' more than 'the earthquake', though socio-scientifically and descriptively correct, is more culturally accurate as 'the ambiguous word expressing our [Haitians']. post-traumatic stupor'. '*Bagay la*' was also more

> indicative of the sense of collective powerlessness that people felt during the very first moments following the earthquake. Not being able to name or define what had happened, while recognizing the full force of its effects individually and collectively, bespoke the sense of utter despair and desolation that immediately gripped the survivors.

The onomatopoeic neologism that represented the noise of the tremors and the crumbling buildings, *goudou goudou*, was later created and has been used since to refer to the January 2010 earthquake. Nancy Dorsinville, 'Goudou Goudou', in *Haiti After the Earthquake*, ed. by Paul Farmer (New York, NY: Public Affairs, 2011), pp. 273–81 (pp. 280–1).

32 Junot Díaz, 'Apocalypse: What Disasters Reveal', *Boston Review*, 1 May 2011, http://bostonreview.net/junot-diaz-apocalypse-haiti-earthquake [accessed 16 December 2017].

33 Anthony Oliver-Smith, 'Haiti's 500-Year Earthquake', in *Tectonic Shifts: Haiti Since the Earthquake*, pp. 18–26 (p. 22).

34 Marc Schuller and Pablo Morales, 'Haiti's Vulnerability to Disasters', in *Tectonic Shifts: Haiti Since the Earthquake*, pp. 11–13 (p. 12).

35 Schuller and Morales, p. 12.

36 Greg Beckett, 'Rethinking the Haitian Crisis', *in The Idea of Haiti: Rethinking Crisis and Development*, ed. by Millery Polyné, pp. 27–51 (p. 41).

37 Beckett, p. 41.

38 Paul Brodwin, *Medicine and Morality in Haiti: The Contest for Healing Power* (Cambridge: Cambridge University Press, 1996), p. 101.

39 Beckett, p. 41.

40 Schuller and Morales do not deny the geological causes of the earthquake – that is the movement of the tectonic plates. Yet the impact these shifts had, the scale of the devastation, is defined by the context in which they occurred, Haiti's socio-economic landscape: 'True, the tectonic plates that shifted along the faults underneath Haiti were by no means products of human action. But to be a *disaster* requires vulnerability.' Schuller and Morales, 'Introduction', p. 4.

41 See James Lewis, *Development in Disaster-Prone Places: Studies of Vulnerability* (London: Intermediate Technology Publications, 1999); Benjamin Goodwin Wisner, Ian Davis, Piers Blaikie and Terry Cannon, *At Risk: Natural Hazards, People's Vulnerability and Disasters* (London: Routledge, 2004); Ben Wisner, J. C. Gaillard and Ilan Kelman, eds., *Handbook of Hazards and Disaster Risk Reduction* (London: Routledge, 2012).

42 Brian Massumi, 'Everywhere You Want to Be: Introduction to Fear', in *The Politics of Everyday Fear*, ed. by Brian Massumi (Minneapolis, MN: University of Minnesota Press, 1993), pp. 3–38 (p. 8).

43 Laënnec Hurbon, 'Catastrophe permanente et reconstruction', *L'Observatoire de la reconstruction*, 6 (2012), pp. 8–10, in Martin Munro, *Writing the Fault Line: Haitian Literature and the Earthquake of 2010* (Liverpool: Liverpool University Press, 2014), p. 6.

44 *Haiti Memory Project* http://haitimemoryproject.org/ [accessed 14 May 2015].

45 Other post-earthquake published responses include survivors' accounts written both by Haitians and non-Haitians (aid and charity workers, members of the rescue forces) as well as narratives of those who visited Haiti in the immediate aftermath of the disaster. Among these are the following works: *Le Tremblement: Haïti, 12 janvier 2010* (2010) by Lionel-Édouard Martin; *Tout Bouge Autour de Moi* (2011) by Dany Laferrière; *Haïti, kenbe la!* (2010) by Rodney Saint-Éloi; *Failles* (2010) by Yanick Lahens; *Belle merveille* by James Noël (2017); *Shaken, Not Stirred: A Survivor's Account of the January 12, 2010 Earthquake in Haiti* (2010) by Jeanne G. Pocius; and *Unshaken: Rising from the Ruins of Haiti's Hotel Montana* (2011) by Dan Woolley. In the second group are Gerard Thomas Straub's *Hidden in the Rubble: A Haitian Pilgrimage to Compassion and Resurrection* (2010); Sandra Marquez Stathis's *Rubble: The Search for a Haitian Boy* (2012); and Ronald J. McMiller Jr's *My 34 Day Memoir of Haiti: From Whence We Come* (2013). In addition, a number of fictional works were published in the wake of the earthquake,

among them Marvin Victor's *Corps mêlés* (2011), Gustave Dabresil's *Récit d'une journée de cauchemar et d'horreur* (2011), and M. Rouzier and J. Rouzier's *38 Seconds: Beyond the Breaking Point* (2011). There were also a few publications aimed specifically at children and young adults: Nick Lake's *In Darkness* (2012), Edwidge Danticat's *Eight Days: A Story of Haiti* (2010), Ann E. Burg's *Serafina's Promise* (2013), and, most recently, Laura Rose Wagner's *Hold Tight, Don't Let Go* (2015). Another sub-category of literary response consists of 'fundraising works' which combine the need for personal engagement with the event and the desire to contribute to relief efforts. One such work is a collection of reflections by Elaine M. Hughes, Patricia J. Koening, Christina L. Ruotolo, Elizabeth B. Thompson and Lynne C. Wigent entitled *The Day the Earth Moved Haiti: From Havoc to Healing* (2011).

46 For an astute analysis of vulnerability as a discourse see Chapter 1, ' "Vulnerability" as western discourse' in Greg Bankoff, *Cultures of Disaster: Society and Natural Hazard in the Philippines* (London and New York: Routledge, 2003), pp. 18–25.

47　　The story of Haiti's earthquake has been told and retold in tens of thousands of blog entries, news stories, You Tube videos, and at least ten English language books. Given the inequalities that marginalize Haiti, particularly the poor majority, the points of view presented to date are dominated by white, foreign do-gooders, either volunteer missions or professional humanitarians. Their stories necessarily celebrate their good intentions and minimize and even denigrated the contribution of Haitians, while also often failing to fully and accurately report the many difficulties that too many Haitians still face.

Schuller and Morales, 'Introduction', pp. 1–11 (p. 3)

48 Toni Pressley-Sanon, 'Haiti: Witnessing as Revolutionary Praxis in Raoul Peck's Films', *Black Camera*, 5 (2013), 34–55 (p. 36).

49 E. Ann Kaplan, *Trauma Culture: The Politics of Terror and Loss in Media and Literature* (New Brunswick, NJ: Rutgers University Press, 2005), p. 123.

50 Roger Luckhurst, *The Trauma Question* (London: Routledge, 2008), p. 15.

51 Terry Rey and Karen Richman, 'The Somatics of Syncretism: Tying Body and Soul in Haitian Religion', *Studies in Religion/Sciences Religieuses*, 39 (3), 379–403 (p. 392).

52 Rey and Richman, p. 392.

53 Karen McCarthy Brown, 'Alourdes: A Case Study of Moral Leadership in Haitian Vodou', in *Saints and Virtues*, ed. by John Stratton Hawley (Berkeley: University of California Press, 1987), pp. 144–67 (p. 152), in Rey and Richman, p. 392.

54 Rey and Richman, p. 392.

55 Rey and Richman, p. 392.

56 For a rigorous analysis of how Vodou communication works through the *pwen*, its connection to possession, as well as a broader, unparalleled in scope discussion of spirit possession see Alessandra Benedicty-Kokken, *Spirit Possession in French, Haitian, and Vodou Thought: An Intellectual History* (Lanham, MD: Lexington Books, 2015).

57 Rey and Richman, p. 392.

58 Nadège T. Clitandre, 'Haitian Exceptionalism in the Caribbean and the Project of Rebuilding Haiti', *Journal of Haitian Studies*, 17 (2011), 146–53 (p. 152).

59 Lowell Juilliard Carr, 'Disaster and the Sequence-Pattern Concept of Social Change', *American Journal of Sociology*, 38 (1932), 207–18 (p. 211).

60 Ilan Kelman, ed., 'Understanding Vulnerability to Understand Disasters' Version 3, 19 January 2009 (Version 1 was 2 September 2007) www.islandvulnerability.org/docs/vulnres.pdf [accessed 12 February 2–18]

61 Nicole Ball, 'The Myth of the Natural Disaster', *Ecologist*, 5 (1975), 368–9.

62 D. J. Blundell, 'Living with Earthquakes', *Disasters*, 1 (1977), 41–6 (p. 41).

63 Ball, pp. 368–9.

64 James Lewis, 'Some Aspects of Disaster Research', *Disasters*, 1 (1977), 241–4 (p. 243).
65 Kenneth Hewitt, 'The idea of calamity in a technocratic age', in *Interpretations of Calamity from the Viewpoint of Human Ecology*, ed. by Kenneth Hewitt (Boston, London, Sydney: Allen & Unwin Inc.: 1983), pp. 3–32 (pp. 10, 14).
66 Lewis, p. 243.
67 See: Ball, Lewis and Michael H. Glantz, 'Nine Fallacies of Natural Disaster: The Case of Sahel', *Climatic Change*, 1 (1977) 69–84.
68 Phil O'Keefe, Ken Westgate and Ben Wisner, 'Toward an Explanation of Disaster Proneness', Occasional Paper No. 1. Disaster Research Unit, University of Bradford (1975), in Ben Wisner, 'Disaster Vulnerability: Scale, Power and Daily Life', *Geo-Journal*, 30 (1993), 127–40 (128); Susan E. Jeffery, 'The Creation of Vulnerability to Natural Disasters: Case Studies from the Dominican Republic', *Disasters* 6 (1982), 38–43.
69 Wisner, 'Disaster Vulnerability', p. 128.
70 See: Hewitt; Glantz and James Lewis, *Development in Disaster-Prone Places: Studies of vulnerability*, (London: Intermediate Technology Publications, 1999); Michael H. Glantz, ed., *Creeping Environmental Problems and Sustainable Development In the Aral Sea Basin* (Cambridge: Cambridge University Press, 1999); Glantz, 'Creeping Environmental Problems', *The World & I* (1994), 218–22; Glantz, 'Creeping environmental phenomena: Are societies equipped to deal with them?', in *Creeping Environmental Phenomena and Societal Responses To Them*, Proceedings of workshop held 7–10 February 1994 in Boulder, Colorado, ed. M. H. Glantz, 1–10. NCAR/ESIG, Boulder, Colorado in Ilan Kelman, J. C Gaillard, Jessica Mercer, 'Climate Change's Role in Disaster Risk Reduction's Future: Beyond Vulnerability and Resilience', *International Journal of Disaster Risk Science*, 6 (2015), 21–7 (p. 25); Anthony Carrigan, 'Towards a Postcolonial Disaster Studies', in *Global Ecologies and the Environmental Humanities: Postcolonial Approaches*, ed. by Elizabeth DeLoughrey, Jill Didur and Anthony Carrigan (New York and London: Routledge, 2015), pp. 117–40.
71 Michael H. Glantz and Dale Jamieson, 'Societal response to Hurricane Mitch and intra-versus intergenerational equity issues: whose norms should apply?', *Risk Analysis*, 20 (2000), 869–82.
72 Bankoff, p. 18.
73 Bankoff, p. 22.
74 Bankoff, p. 22; see also Kenneth Hewitt, *Regions of Risk: A Geographical Introduction to Disaster* (Harlow: Longman, 1997).
75 John McLeod, 'Introduction: Postcolonial Environments', *The Journal of Commonwealth Literature*, 51 (2016), 192–5 (p. 194).
76 However helpful to convey that sense of non-spectacular yet consistently hurtful violence, Rob Nixon's formulation of slow violence, in its focus on 'the poor' and the 'Global South', risks overlooking the reality of uneven vulnerabilities even within these groups as well as the fact that the violence of multi-scalar vulnerability is not reducible to poverty. Rob Nixon, *Slow Violence and the Environmentalism of the Poor* (Cambridge, MA: Harvard University Press, 2011), p. 2.
77 Nixon, p. 2.
78 Oliver-Smith, *'Haiti's 500-Year Earthquake'*, p. 18.
79 Joyelle McSweeney, 'Poetics, Revelations, and Catastrophes: an Interview with Kamau Brathwaite', *Rain Taxi*, Fall (2005), www.raintaxi.com/poetics-revelations-and-catastrophes-an-interview-with-kamau-brathwaite/ [accessed 26 September 2015].
80 Oliver-Smith, p. 18
81 For a comprehensive and up-to-date overview of postcolonial studies, see: Graham Huggan, ed., *The Oxford Handbook of Postcolonial Studies* (Oxford: Oxford University Press, 2013); John McLeod, ed., *The Routledge Companion to Postcolonial Studies* (London: Routledge, 2007).

82 Ato Quayson, *Postcolonialism: Theory, Practice or Process?* (Cambridge: Polity, 2000), p. 2.

83 Quayson, p. 9.

84 Quayson, p. 11.

85 Gina Athena Ulysse, *Why Haiti Needs New Narratives: A Post-Quake Chronicle* (Middletown, CT: Wesleyan University Press, 2015).

86 Amy E. Potter, 'Voodoo, Zombies, and Mermaids', p. 208. Potter's analysis provides a study of 700 articles on Haiti in six major American newspapers in 2004. Robert Lawless's book is the key critical study of media portrayals of Haiti. It analyses disconnect between journalistic, ethnographic and scholarly ways of seeing Haiti – a prevailing dichotomy. See Robert Lawless, *Haiti's Bad Press* (Rochester, VT: Schenkman Books, 1992). Bob Corbett's website database of the same title, *Haiti's Bad Press*, takes up the challenge set by Lawless's work, 20 years after its publication, by providing a sample of recent media portrayals of Haiti which reinforce damaging stereotypes of the island. See: Bob Corbett, *Haiti's Bad Press* http://haitibadpress.tumblr.com/ [accessed 25 February 2014]. For an analysis of specifically post-earthquake media response, many of which are just reiterations of earlier tropes and denigrating imagery, see also Patrick Bellegarde-Smith, 'A Man-Made Disaster: The Earthquake of January 12, 2010 – A Haitian Perspective', *Journal of Black Studies*, 42 (2011), 264–75.

87 Christopher Winks, 'Martin Munro: *Tropical Apocalypse: Haiti and the Caribbean End Times* (Charlottesville: University of Virginia Press, 2015) [book review]', *New West Indian Guide/Nieuwe West-Indische Gids*, 91 (2017), 149.

88 Martin Munro, *Tropical Apocalypse: Haiti and the Caribbean End Times* (Charlottesville: University of Virginia Press, 2015), p. 199.

89 Beckett, p. 27.

90 Kaiama L. Glover, 'New Narratives of Haiti', p. 198.

91 Millery Polyné, 'To Make Visible the Invisible Epistemological Order: Haiti, Singularity, and Newness', in *Idea of Haiti*, pp. xi–xxvii (p. xxvi).

92 Carrigan, 'Towards a Postcolonial Disaster Studies', p. 134.

93 Beckett, p. 29.

94 Ulysse, *Why Haiti Needs New Narratives*, p. 16.

95 Ulysse, p. 16.

96 Mark D. Anderson, *Disaster Writing: The Cultural Politics of Catastrophe in Latin America* (Charlottesville, VA. and London: University of Virginia Press, 2011).

97 Anderson, p. 2.

98 Whereas Rachel Douglas's discussion of the relationship between archiving and rewriting in post-earthquake writing focuses, like Munro, on Haitian literature, my analysis extends this notion of an open-ended archive of the earthquake to include other narratives of those thrust into the disaster experience. Rachel Douglas, 'Writing the Haitian Earthquake and Creating Archives', *Continents manuscrits*, 8 (2017) http://journals.openedition.org/coma/859 [accessed 11 February 2018], p. 12.

99 'In the end, the triumphant version of events, often formulated through consensus with various competing accounts, achieves canonical status as the basis for political action, and frequently, but not always, it persists as memory in the popular imagination.' Anderson, p. 7.

100 Anderson, p. 28.

101 Yarimar Bonilla, *Non-Sovereign Futures: French Caribbean Politics in the Wake of Disenchantment* (Chicago and London: University of Chicago Press, 2015), p. xiv.

102 Anderson, p. 2.

103 Gilbert F. White, *Natural Hazards: Local, National, Global* (Oxford: Oxford University Press, 1974), in James Lewis, 'Some Aspects of Disaster Research', *Disasters*, 1 (1977), 241–4 (p. 243).

104 Rodney Saint-Éloi, *Haïti, kenbe la! 35 secondes et mon pays à reconstruire* (Neuilly-sur-Seine: Éditions Michel Lafon, 2010), p. 38.

105 Edward Kamau Brathwaite, *History of the Voice: The Development of Nation Language in Anglophone Caribbean Poetry* (London: New Beacon, 1984), p. 265.

106 Martin Munro, *Haiti Rising: Haitian History, Culture and the Earthquake of 2010* (Liverpool: Liverpool University Press, 2010).

107 Pierre Beaudelaine and Nataša Ďurovičová, eds., *How to Write an Earthquake: Comment écrire et quoi écrire/Mo pou 12 janvye* (Iowa City, IA: Autumn Hill Books, 2011). An initiative of the International Writing Program in Iowa (US), the anthology brings together the early reactions to the January 2010 earthquake by a dozen Haitian poets and writers – a category that encompasses the three linguistic realities of Haitian writing in Kreyòl, French and English. This emphasis on the importance of braiding together the three languages directly shapes the form of the anthology. The 'edition gives priority to the original texts. The English translations then follow in the second part of the volume. In the e-edition of this book this matter is, fortunately, moot, as the reader can herself choose the reading sequence' (*How to Write an Earthquake*, Editor's Note). Published in 2011, the project started the day after the earthquake and has at its core 'the ethical question that each writer, each poet, each artist has confronted in the face of overwhelming catastrophes: what can I say? Of what use is writing?' Joëlle Vitiello, 'Introduction', p. 3. The authors included, and arranged in alphabetical order in the anthology, are: Raoul Altidor; Bonel Auguste; Chenald Augustin; Dominique Batraville; Edwidge Danticat; Valerie Deus; Gaspard Dorelien; Forteston 'Lokandya' Fenelon; Yanick Lahens, Joël Lorquet; Lucie Carmel Paul-Austin; Beaudelaine Pierre; Claude Bernard Sérant; Patrick Sylvain; Lyonel Trouillot; Joujou Turenne.

108 Beaudelaine and Ďurovičová, p. 3.

109 Beaudelaine and Ďurovičová, p. 6.

110 'We are here! We are united together!' Beaudelaine and Ďurovičová, p. 6.

111 Martin Munro, *Writing on the Fault Line: Haitian Literature and the Earthquake of 2010* (Liverpool: Liverpool University Press, 2014), p. 2.

112 Vanborre also sees the ability to create a space of reflection as the main task and value of literary responses to the disaster:

> Nous ne pouvons pas effleurer la réalité du tremblement de terre, de la catastrophe, de la souffrance et de la mort. Mais la littérature offre un moyen de réfléchir, de penser, sans prétendre nous offrir la réalité de l'expérience dont nous n'avons justement pas fait l'expérience. Il est impossible de se souvenir de la réalité exacte, de l'expérience, il est également impossible de ne pas se souvenir, et la littérature seule peut maintenir cette tension impossible et nécessaire.

> We cannot touch the reality of the earthquake, catastrophe, suffering, and death. But literature gives the opportunity to reflect and think without pretending to offer the reality of an experience which we had not experienced. It is impossible to remember the exact reality or the experience, it is also impossible not to remember, and only literature can maintain this impossible and necessary tension.
>
>   Emmanuelle Anne Vanborre, 'Haïti après le tremblement de terre: le témoignage impossible et nécessaire', in *Haïti après le tremblement de terre: la forme, le rôle et le pouvoir de l' écriture* (New York, NY: Peter Lang, 2014), pp. 3–14 (p. 12)

113 Vanborre, p. 12.

114 Munro, *Writing on the Fault Line*, pp. 2–3.

115 Munro, p. 3.

116 'Témoignagne', in *Reverso: Collins Online English-French Dictionary* (HarperCollinsPublishers,2012)http://dictionary.reverso.net/french-english/t%C3%A9moignage [accessed 26 May 2015].

117   The testimonial account is therefore composed of opposing and irreconcilable elements. It contains the possibility of fiction while eliminating it in order to assume its status as a testimony anchored in an account of a witness and in some lived and recounted reality.

Vanborre, p. 4

118 The repeated use of these images is voyeuristic at best, as Sibylle Fischer remarks, creating an image of Haiti as *incomprehensible*. These anonymous bodies exposed to our gaze are without name, history or identity:

> dead black bodies, wherever you look. People without names, without history, without location: mere bodies, all black, all shoveled into mass graves without much ado. So different from our protective sense of bodily integrity in the North; yet familiar, since it is Haiti: exposed to a gaze which at times borders on the pornographic, a country up for grabs. Sibylle Fischer, 'Beyond Comprehension,
> *Social Text: Ayiti Kraze/Haiti in Fragments*, 26 January. 2010, http://
> socialtextjournal.org/periscope_article/beyond_comprehension/
> [accessed 01 August 2014]

119   Some Haitian intellectuals sought to combat those images but in so doing they inadvertently revealed their complicity not only in the negative discursive construction of their country, but also in the economic and military re-colonisation of Haiti over the last decade.

> Andrew Leak, 'A Vain Fascination: Writing from and about Haiti after the
> Earthquake', *Bulletin of Latin American Research*, 32 (2013) 394–406 (p. 394)

120 'Je t'écris Haïti', *Libération*, 19 January 2010 www.etonnants-oyageurs.com/spip.php?article4894 [accessed 19 February 2018].

2 Disaster's past, present and future

Beyond exposure, failure and resilience

Narratives of the January disaster experience, not a singular event but rather a chain of overwhelming events and an 'irreducible mix of coexisting planes of experience',[1] are embedded and entangled in these longer histories that preceded and shaped the ways in which the disaster unfolded. Haiti's environmental and political histories are two such key threads. They further reaffirm the disaster's compound and accretive character and demonstrate the need to conceptualise the processual character of the January 2010 disaster beyond the confining constructions of Haiti in terms of its unfortunate exposure, failing and failed state with the population's resilience portrayed as the only possible antidote to this seemingly unalterable situation.[2] In addition, these longer histories clearly show that the 2010 disaster was not a 'unique' occurrence. Rather, the event, and what followed upon it, was underpinned by systemic violence, ongoing political and economic marginalisation, and specific histories of environmental exploitation and degradation in a particular – and rendered particularly vulnerable – part of the world.

Socio-environmental histories and topographies: beyond exposure

Haiti's geographical location and geological formation, the island's exposure to hazards, is only one, external dimension of the local community's and individuals' susceptibility to harm.[3] The island is located directly on the path of tropical storms and is consequently exposed to flooding, heavy rainfalls, landslides, aggravated as a result of unsustainable urban planning and construction practices; for example, in zones that have a high level of hazard.[4] The geological make-up of the island is the other dimension of this exposure. Scientists have identified and located at least four major fault lines in Haiti – the Enriquillo-Plantain Garden Fault, the Septentrional Fault, the Trans-Haitian Fault and the Léogâne Fault[5] – environmental hazards that are rendered fatal through compound vulnerability. For example, the 1751 earthquake destroyed colonial Port-au-Prince. Almost a century later, in 1842, magnitude Mw8.1 tremors hit the northern cities of Cap Haitian and Port-de-Paix, triggering also a tsunami.[6] Supplementing first analyses of the 2010 tremors, later research demonstrated that the January earthquake did not just involve the straightforward accommodation of oblique relative motion between the Caribbean and North American plates along the Enriquillo-Plantain Garden fault zone. Rather, as Gavin P. Hayes and others argue, the rupture process 'involved slip on multiple faults',[7] and only the 'primary surface deformation was driven by rupture on blind thrust faults with only minor, deep, lateral slip along or near the main Enriquillo-Plantain Garden fault zone'.[8] It is highly likely that the remaining 'shallow shear strain will be released in future surface-rupturing earthquakes on the Enriquillo-Plantain Garden fault zone'[9] as it has already occurred 'in inferred Holocene and probable historic events'.[10]

The geological features and the make-up of the island, often viewed as at once unchangeable and unalterable, are all the more important in the context of increasing urbanisation in Haiti, uneven population distribution and density across the country – urban factors that add to and can determine the hazardous and potentially deadly nature of environmental phenomena. First, Haiti's capital is located near or directly over two active fault systems.[11] Second, it is the most densely populated and urbanised area in the country. As of 2015 approximately 2.5 million inhabitants, out of a population of approximately 10.6 million (2017 estimates),[12] live in Port-au-Prince. The highly concentrated downtown areas are home to those who can only afford the least desirable land with the 'low-density hilly areas of high amenity with rich vegetation'[13] occupied by the economically privileged. The continuing growth of urban neighbourhoods, which tend to occupy marginal land, without an accompanying development of the city's infrastructure further contributes to people's vulnerability in case of flood, mudslide or earthquake. These marginalised neighbourhoods are disproportionately affected, e.g. through the chronic lack of building regulations and adequate construction material, while the risk of potential epidemics is much greater in areas with a high concentration of people and lack of access to clean water.[14] For these communities, each seasonal flood or even an increase of rainfall, not to mention seismic tremors, can become a catastrophic disaster.

Together, these factors converge with long-term, historical, socio-economic and political challenges – high unemployment, growing inflation, high level of reliance on international aid and foreign investment – to create the setting of compound and multi-scalar vulnerability that transformed the 2010 earthquake into a disaster and that continues to pose a threat to Haiti's populations. Or as Harley F. Etienne powerfully puts it:

> the legacy of French colonialism, US imperialism, and unstable leadership has starved Haiti of the planning framework [and] [t]he vestiges of this starvation have led to an urban crisis of sorts that defines the urban form of Haiti's major cities.[15]

In the immediate aftermath of the event, not only the city's topography but also one's previous socio-economic network and standing played a decisive role, providing or limiting 'access to shelter-related resources for those with connection'.[16] In effect, this 'accentuate[d] pre-existing inequalities or create[d] new inequalities among displaced Haitians',[17] further reaffirming the constructed and sustained character of vulnerability of marginalised urban populations. In sum, the city's make-up – the differentiated exposure to other recurring hazards such as landslides, flash floods or coastal surge – and the resulting potential for recovery are both shaped by the drastically uneven economic and political status of its inhabitants.

Moreover, the highly asymmetrical socio-economic topography of Port-au-Prince is directly linked to increasing migration from provinces to the capital driven by, long before January 2010, centralisation of key resources (e.g. university education) and political power in Port-au-Prince as well as compound challenges – including soil erosion, deforestation, marginalisation and damaging food imports – faced by Haitian farmers. Individual decisions to move to the city, whether as a final destination or a stop-over before further emigration abroad, however, are not reducible to a tally of 'push' and 'pull' factors. One's choice to migrate has a complex causation, which, as Frank Graziano demonstrates, entails 'the interactions of individual disposition, cultural context, and domestic and global conditions',[18] and is often rooted in unquantifiable, sometimes 'arational',[19] personal resolve, one's 'aspirations, responsibility, desperation, beliefs'[20] as well as the sense of there being no other option available than to migrate in order to support one's livelihood. This complex layering is equally important to one's subsequent decision whether to leave or stay or where to go, not least following disastrous events like the 2010 earthquake.

In response to the scale of devastation after the January tremors, some analyses hoped to alleviate the socio-environmental pressures faced by the inhabitants of Port-au-Prince and suggested moving Haiti's capital to seismologically safer areas of the country.[21] Others called for the reconstruction process to be shaped accordingly with the country's known geological features, arguing that the building of high-stake infrastructure (e.g. hospitals, schools, relief centres) in the vicinity of active fault should be prohibited.[22] Most affected in terms of casualties, Port-au-Prince, however, was not the only city or region confronted with the force of the tremors. As mentioned before, 80 per cent of Léogâne, a town west of Port-au-Prince and closest to the earthquake's epicentre, was destroyed.[23] Jacmel, the cultural capital in the south of Haiti, famous for its annual carnival celebrations, was also affected with hundreds of deaths reported and damage to many of its historic buildings. Outside of Haiti, the tremors were felt throughout 'the Dominican Republic as well as in parts of nearby Cuba, Jamaica, and Puerto Rico';[24] some informal reports testify to the tremors being felt as far as Caracas in Venezuela.[25] Thus, the 2010 earthquake and the disaster it had become cannot be contained and reduced to one site, Port-au Prince. Likewise, as narratives of the earthquake make clear (see especially Part III), even this one specific city is not reducible to its physical topography, layout or spatial features that can be easily adjusted. Rather, it is a layered and palimpsestic place, of cultural, historical and community significance, and a complex site 'on which [one's] subjectivity is founded'.[26]

Consequently, both the material destruction of the Haitian capital, the nearby Léogâne and other affected neighbourhoods and towns including Kenscoff and Jacmel, as well as post-earthquake reconstruction of it – whether it entails temporary displacement of local inhabitants outside of the city or to a different part of it – are equally radical reconfigurations of non-material community networks, family and affective relationships. All are also marked by a sense of profound loss. In the complex aftermath of the 2010 tremors, it becomes clear that post-earthquake reconstruction and recovery need to account for and take consideration of such composite make-up and the many non-material meanings of urban space. Or, as Dany Laferrière (Part III) pointedly expresses: 'The earthquake didn't destroy Port-au-Prince; no one can build a new city without thinking of the old. The human landscape counts. Its memory will link the old and the new. Nothing is every begun from scratch.'[27]

Political history: beyond failure

These intertwined histories of urban topography, environmental pressures, migration and displacement were directly linked to Haiti's wider political history, defined to a great extent by internal instability and foreign intervention and experienced, after Erica Caple James, as *ensekirite* (insecurity in Haitian Kreyòl); that is, 'an ontological, pernicious, and powerful state of "routinized ruptures" that plagues Haiti's efforts to consolidate its democracy and create a climate of peace and security'.[28] These histories of quotidian ruptures and long-term depravations also shaped the ways in which the post-disaster relief and reconstruction efforts have been envisaged and practised, at the exclusion of the Haitian majority, since the disaster struck. This combination of external interference, internal unrest and structures of marginalisation and exclusion is a crucial determinant of the country's compound vulnerability, made even more apparent immediately after the 2010 earthquake. At the same time, vulnerability is not simply synonymous with and reducible to 'poverty', a label that as Alessandra Benedicty-Kokken demonstrates is still 'based on a civilizing mission, dividing nations into those that are "developed" and those that aren't'.[29] Soon after the earthquake, the then head of the UN's International Strategy for Disaster Reduction, Salvano Briceňo, suggested a causal relationship between economic poverty and Haiti's susceptibility to environmental hazards. In a comment made on 13 January 2010, he said: 'The major problem of course is that Haiti is extremely poor and it's poverty that is at the core of all of these disasters.'[30] Whereas economic marginalisation and lack of access to resources is an important aspect of vulnerability, it is not the sole determinant of it. Rather, it is the combination of long-term factors and processes, compound and across multiple scales, that, together, create the vulnerability of a given place. Although such single-issued focused approaches suggest, with a lot of hope and an equal amount of ignorance, that a solution to the 'disaster-problem' is indeed easily reachable, by not addressing the multi-layered character of vulnerability – that includes environmental issues, structures of governance, foreign interference or insufficient regulations (e.g. building codes) – they effectively do the opposite, namely increase a community's susceptibility to environmental hazards. Finally, an over-emphasis on some of the economic factors which can contribute to, but do not necessarily determine in a casual way, a country's vulnerability fails to take into account its uneven and multifarious, processual character and the plethora of quotidian manifestations vulnerability takes.

In contrast to such limited and limiting interpretations of vulnerability that, in effect, also confine post-earthquake aid and recovery to a trajectory of economic 'progress', a longer, more comprehensive view exposes historical determinants of vulnerability, at once constructed and sustained. Regardless of their varying vantage points, narratives discussed in the book testify to this multi-layered and uneven experience of vulnerability – whether seen in highly personal terms or in direct relation to issues of foreign presence, relief and reconstruction practices – and the longer chronologies of marginalisation and exclusion of the majority of the Haitian population in which they are entangled. To this end, the selected accounts reference, implicitly or explicitly, or embed directly within their narrative structure (e.g. Part IV) key events from across Haiti's history, from the colonial period to the Haitian Revolution, twentieth-century dictatorships and the subsequent, and the ongoing struggle for a democratic and fully-sovereign state.

In so doing, these narratives also point to the processual character of the event which, despite its scale, is very much embedded in rather than disconnected from national, regional and global pasts. Against instant depictions of Haiti as 'marginal' and 'unique', the country's recent and distant past offers reminders that Haiti's politics and history have long been enmeshed within the wider Atlantic and world history, defined by the violence of slavery and of the Middle Passage, but, equally, by radical revolutionary and successful emancipatory struggles. Consequently, Haiti and the 2010 earthquake are not exceptions that stand outside time and history, but 'rather [have] a history deeply intertwined with other national histories, a history that is an exemplar of many modern historical processes'.[31]

The events evoked across the analysed texts stretch across the last five centuries of Haiti's history from Columbus's landing on the northern coast of the island to very recent history of post-earthquake aid and the cholera epidemic. These include (in chronological order): Columbus's 1492 'discovery' (Saint-Éloi) and its colonial aftermath; the 1791–1804 Revolution (Lake); the U.S.-backed father-son dictatorships of François 'Papa Doc' and Jean-Claude 'Baby Doc' Duvalier (1957–86); the early 1990s and the ascension to power of Jean-Bertrand Aristide (Lake and Marquez Stathis) as well as his controversial and still-hotly debated departure; the subsequent years of instability and UN involvement (Lake and Marquez Stathis); the experience of diasporic displacement (Saint-Éloi, Laferrière); and the post-earthquake failings of aid and the cholera epidemic (Laferrière, Wagner). In no way an exhaustive list, these historical references highlight some of the key moments in the formation of the Haitian state, a process which is inseparable from foreign interference and intervention and has global reverberations. Finally, these historical pointers, as well as the ways in which they are used and embedded within the different narratives, can further serve to demarcate the narrator's position, especially in non-fiction and autobiographical accounts, while testifying to the varying frames of reference and interpretative frameworks which are mobilised in order to account for the January disaster, its history and its impact.

The Haitian diasporic writer and earthquake survivor Rodney Saint-Éloi (Part II), for example, evokes the 1492 'discoveries' as part of his conceptualisation of colonial reverberations still audible and detectible in the 2010 disaster using, what he sees as, the first earthquake in the country's history in order to question and imagine anew a collective, and open-ended archive of Haiti's more recent disaster. Nick Lake's novel (Part IV), a fictional non-autobiographical account, turns to a later, epochal event, a radical undoing of colonial rule, namely the 1791–1804 Revolution. A subject of increasing scholarly interest,[32] the Revolution is a founding national event, and 'one of the key moments in the history of human emancipation'.[33] In the text, it functions as a central conceptual metaphor that allows the narrator to suggest, with varying degrees of success, visions of post-earthquake personal and collective recovery as necessarily rooted in Haiti's emancipatory struggle. By creatively charting the beginnings and the course of the Haitian Revolution, from the Bois Caïman (Bwa Kayiman in Haitian Kreyòl) ceremony to Toussaint Louverture's imprisonment,[34] and by establishing parallels between the life of the revolutionary general and that of a young boy from Cité Soleil, one of Port-au-Prince's many informal neighbourhoods, the text voices a pointed critique of disempowering and further marginalising politics of aid and practices of foreign intervention in Haiti. To this end, the fictional narrative also interweaves the young boy's personal reflections as well as recollected family stories from the time of growing popularity, rise to power and first term of Jean-Bertrand Aristide, two centuries after Louverture's ascent.

Following almost three decades of the Duvaliers' reign of terror, in 1990, Jean-Bertrand Aristide's victory in Haiti's first free elections inspired a wave of optimism and hope that, after years of oppression, Haiti-led change might finally be possible. A former priest and a long-time opposition activist, Aristide enjoyed unprecedented levels of popular backing and his electorate included 'peasant organizations and the urban poor, together with progressive members of the church and the liberal elite'.[35] The young narrator in Lake's novel partakes in and recalls some of that early enthusiasm and wide-ranging support that swept across the earlier excluded popular neighbourhoods. In a gesture of creative reimaging of a historic moment, Shorty is put at the centre of these events: his mom gives birth to him during one of Aristide's charismatic sermons with the priest safely delivering and blessing the boy.

However, eight months after the election, a coup led by General Raoul Cédras forced Aristide into exile. Aristide conducted his lobbying campaign from exile and his supporters in Haiti and abroad called for an international intervention to help reinstate him. Finally, in 1994, following a series of unsuccessful negotiations with the coup leaders, then President Bill Clinton sent US Marines to Haiti with Aristide on board.[36] This help was not unconditional: Aristide had to commit to a number of neoliberal market policies which contradicted his earlier socially oriented political and economic programme. At the same time, he was highly aware that Haiti's position was a precarious one and that the country could not afford 'to plan a political economy that would turn the entire world against it'.[37] The choice for Aristide seemed clear: 'either we enter a global economic system in which we know we cannot survive, or, we refuse, and face death by slow starvation'.[38] The U.S. and international financial organisations applied further pressure to privatise state assets, and Aristide was left with little room for manoeuvre. One thing he insisted on carrying out, despite US opposition, was the full demobilisation of the army. After he completed his term in 1996, Aristide ceded power to his supporter René Préval, who remained in office until 2001. In 2001, Aristide was re-elected after a vote that many members of the international community considered rigged.[39] The anti-Aristide opposition, too, denounced the electoral process as undemocratic and illegitimate. The following years were marked by increasing political stalemate and Aristide faced growing accusations of allowing or even encouraging state-sanctioned violence against opponents. Indeed, the use and extent of violence during Aristide's regime continues to be one of the most hotly debated aspects of his rule.[40] Equally, his second departure in 2004, when he was escorted from Haiti by US Marines, remains a point of controversy, with contrasting interpretations of the event as either a voluntary departure, a kidnapping or a coup. These unresolved tensions, the interwoven political interests and hemispheric alliances, made all the more clear during Aristide's term and in the years and decades following his presidencies, determine the life of the young protagonist in Lake's novel, the boy's place and sense of the world before the earthquake struck, and, after his rescue, shape his vision of what is possible for him and his family in this completely changed Haiti.

Not only Lake's *In Darkness* but also Marquez Stathis's *Rubble* directly refer to these events of the decade and the controversies surrounding Aristide's presidency, viewing them from the perspective of a sceptical American observer (*Rubble*) and a young boy (*In Darkness*) for whom Aristide is a legendary figure and the only contemporary leader who put Haiti's excluded majority at the heart of a radical political programme. Marquez Stathis (Part III), while openly admitting her own sense of confusion and inability to effectively negotiate her position as an OAS observer, recalls extensively her first journeys and stays in Haiti in the 1990s in order to set her own position and express her investment in the country. This then serves as a means of establishing the narrative's authority and that of the seemingly objective pronouncements, addressed primarily to an American audience, the text makes on the country's past and present, before and after the 2010 earthquake.

Dany Laferrière, for his part, builds his memory space around the varying sites, built environments and topography of the city (Part III), their changing personal and collective significance over the years, from his early memories of

Haiti, its current marginalisation and vulnerability to a consideration of the country's uncertain futures. Laferrière's literary journey is at the same time a part of his wider attempt to come to terms with his responsibility as a diasporic writer, a highly privileged and demanding, liminal position: Laferrière left Haiti in 1978, during Baby Doc's regime (following an assassination of his friend by Tonton Macoutes).[41] Ideas of diasporic belonging and the difficult coupling of the deep emotional proximity and connection and the reality of physical distance and detachment are also central to Laura Rose Wagner's young adult novel (Part IV). Here, the text turns to immediate and ongoing post-earthquake crises – such as the increasingly evident permanence of Internally Displaced People's camps, the UN-caused cholera epidemic,[42] among others – to think through notions of incomplete recovery and unhealed futures, in split times; that is the reality of separated families and parallel lives. Consequently, these longer histories of the January tremors challenge ideas of recovery and reconstruction as a return to the pre-disaster state, reframing the disaster as, at once, a violent manifestation and a coming together of concurring, interlocked and ongoing processes of vulnerability formation.

The abundance of historical markers in these texts is somewhat countered by a comparable lack of direct historical references in both Martin's memoir (Part II) and Woolley's conversation narrative (Part IV). Here, the immediacy of the experience of the disaster, formal focalisation and aesthetic features – including language, imagery, symbolism and pace of the narrative – of these two texts shift the emphasis towards the emotional and, in Woolley's case, spiritual frames of reference and interpretation. Yet history is not absent even from these highly personal accounts. The very circumstances of the authors' respective presence in Haiti – Martin's longstanding commitment to Haiti and his status as a French author living in Martinique and Woolley's almost accidental journey to Haiti as part of a US charity – are inseparable from and deeply embedded within the history of foreign presence in Haiti. In sum, the narratives analysed here repeatedly point to the multi-layered constructions and differentiated lived reality of Haiti's vulnerability and underline the connections between the country's environmental and political history and the personal and collective experience of the 2010 disaster.

These aforementioned events constitute a brief pre-history of the disaster – the material contexts that precipitated it and that shaped its aftermath. This material *history* cannot be divorced from the considerations of *historiography*: how events are written and remembered. Despite growing scholarly interest in and public awareness of Haiti's global significance,[43] the country's post-revolutionary history remains largely unknown or is often seen as a series of disasters, dictatorships and ever-growing poverty that followed the heroic, but ultimately fatal, revolutionary impulse. However, neither history nor historiography can be reduced to discourse and discursive practices. The historical excavation of the earthquake therefore has two dimensions. First, it reveals a number of processes and dynamics which, to varying degrees, would condition the scale, impact and specificity of the earthquake; and second, it interrogates the different ways in which, and ends to which, the past is narrativised and remembered as a whole. In the immediate aftermath of the disaster, this dual task of contextualisation and interrogation would become ever more urgent.

Despite an increasing realisation of the many socio-economic fault lines, as well as the role of foreign intervention in determining the scale of the events and their aftermath, the rhetorical framings of the earthquake, and of Haiti's present predicament more generally, by Pat Robertson, David Brooks and others,[44] have not substantially changed. Instead, they have tended to cite previous tropes of failure, vulnerability, progress-resistant culture and lack of 'political culture' as reasons behind the country's current parlous position. These hasty configurations mask neocolonial and neoliberal interpretative frames which fix Haiti in a position of dependence on external assistance and intervention. In addition, such re-inscription of Haiti as a failed (and continually failing) state positions its 'glorious past' in radical contrast to its 'doomed present'. In this binary framework, international interventions – whether humanitarian, economic or political – have been presented as attempts to 'save' Haiti against all odds. Such easily rehearsed rhetorical portrayals depict Haiti as a place of multiple colliding realities, no longer governed by a set of rules, and as a site outside of global modernity and the processes which have shaped it.

Aware of the ways in which Haiti's history and culture have been subject to effacement, Haitian Studies scholars have long fought against these problematic narratives of Haiti which, as Nadège Clitandre shows, were uncritically reaffirmed in response to the disaster:

> While the media maintained its focus on the narrative of Haiti as impoverished country through images of naked and wounded black bodies, Haitian scholars and scholars of Haitian Studies were quick to intervene to expose the ways in which this narrative of degradation presented by the media effaces a long history. That is the history of enslavement, involuntary migration and displacement; the history of colonialism, foreign intervention, forced isolation, and economic exploitation.[45]

Here, the tensions between competing media and scholarly narratives are made explicit. The complexity of the latter is generally set against the ignorance of the former. Fatton, Katz, Bellegarde-Smith, Asante[46] and Lundy, among others,[47] have all underlined the need to consider the multiple contexts for the event, its long history, as well as for the differentiated responses it elicited. In so doing these scholars, and with them this book, oppose the dominant rhetoric deployed to account for the scale of the event which was a reiteration of earlier claims of ignorance, paired with and translated into practices of ignoring and marginalising Haiti in political, epistemological and discursive terms – a compound erasure and conscious ignorance of external factors and processes which have shaped and connect local, regional and hemispheric histories and politics.

Remnant dwelling: beyond resilience

Whereas vulnerability is used to denote a community's or an individual's propensity to be harmed, resilience is often positioned as its stark opposite with post-disaster humanitarian aid hoping to 'build resilience' of affected communities. Yet, although they do have some converse characteristics, vulnerability and resilience are not the exact opposite and can exist simultaneously.[48] As a critical term, resilience is employed in fields as diverse as medicine, psychology, engineering and ecology.[49] It is applied to a wide range of contexts from disaster risk reduction,[50] individual responses to disasters, to community capacity, cultural resources, ecosystems capacity,[51] being used to describe national and international structures of support and adaptation, or lack thereof, as well as processes of individual adaptation and 'behaviors, thoughts and actions that can be learned and developed in anyone'.[52] Across the many uses of the concept,[53] resilience emerges, conversely, 'as a trait, a process, or an outcome'.[54] These wide-ranging application and differentiated understandings of it[55] suggest that resilience, an ever more elusive designation, might be, after Luckhurst, one example of such heterogeneous conceptual 'knot',[56] an assembly of ideas and practices tied together within one term.

Yet despite the term's seeming flexibility and transferability, the concept of resilience is, too, a form of narrative and a discourse. Thus, it can effectively conceal the many ways in which communities, for example indigenous groups rooted in distinct worldviews and linguistic practices,[57] and individuals respond to and live through moments of crises as well as their unfolding, disastrous histories. The term, whether functioning as an aim or a characteristic of a group, can act as an exploitative and marginalising tool and a dehumanising discourse. Under the impression of bouncing back, as a recovery from hazards[58] or a return to the pre-disaster normal, it detracts from 'focusing on individual, community-level, and covert long-term infrastructure disorders that must be addressed',[59] effectively foreclosing any attempts to counter or amend structures of vulnerability.

In the months following upon the January disaster, the limitations of resilience, whether as a desirable future goal or an imposed category shaping practices of aid, became even more obvious. The still-repeated claims of Haitians' 'extraordinary resilience' and concomitant narrative of the country's and the event's uniqueness, as Kaiama L. Glover powerfully demonstrates, were 'premised not only on the notion of Haiti's endless suffering but also on a concomitant notion of the Haitian people's endless *capacity* for suffering' [italics original].[60] This insistence on Haiti's exceptionalism, whether used positively to counter the island's marginalisation or employed in order to underline its otherness, ultimately 'conflates the *super*- and the *sub*human'.[61] It positions Haiti and its people 'at the extreme poles of the human condition – to the point where Haitians are excluded from existence and, to an extent, from consideration within the borders of recognizable and lovable (empathy-inducing) humanity'.[62] In short, Haiti still functions as 'the Other', as a site of unbridgeable and incomprehensible difference. Individual and collective agency seems unattainable.

In this configuration, the underlying reasons for the particularity of Haiti's history – not least centuries of foreign intervention, ongoing ecological challenges and internal instability linked to international economic policies – remain unacknowledged. As a result, these depictions – which still echo in some of the texts analysed here, reaffirming a vision of Haiti as an antithesis to all that is orderly and cultured – risk contributing to racialised discourses on the country's 'backwardness' that are then translated into 'benevolent' policies of aid and practices of international development. If unqualified, such claims of 'natural resilience' can dangerously imply some sort of 'innate' ability on the part of Haitians to cope with disasters while confirming that, whatever the future might bring, Haitians will somehow 'rise above it all':

> While the Haitian state and economic elite are condemned, excoriated, and reviled, the nonelite Haitian population is consistently heralded for its exceptional, superhuman ability to withstand. To be bullied. To be displaced and disenfranchised.[63]

Although one might argue that Haitians have indeed faced 'many political, economic and environmental storms over the centuries',[64] such all-purpose statements belie the complex processes and causes behind each particular disaster and the structures of vulnerability that underpin it.

In effect, 'resilience', and its blanket use as an unquestionable goal or as an assumed characteristic, can feature almost as a remedy for the vulnerability of island populations such as Haiti's. Effectively, it undermines the urgency and the real needs of the local population and, what Edwidge Danticat terms, the 'passive hurt'[65] Haitians are subjected to, namely the lack of urgency on the part of the international community, long after the world's attention has turned to other disasters. Rather than speaking of Haitians' resilience, which gives it an ennobling dimension and reaffirms discourses of Haitian exceptionalism, Haitians' responses to the immediate aftermath of disaster and its long and violent history, in the words of Pascale Monnin, are better understood as '*survenance/survivance*'.[66] The French term, used by the Haitian Swiss painter, pairs 'survenance' (emergence) with 'survivance' (survival), describes one's dizzying and overwhelming confrontation with the all-encompassing destruction and captures the dual attempt to be and to reinstate one's presence and to form a mode of emerging and emergent and intense survival which, in the philosopher Jacques Derrida's words, 'is neither life nor death pure and simple, a sense that is not thinkable on the basis of the opposition between life and death'.[67] Finally, '*survenance/survivance*', as a non-binary concept includes the ideas of remnant presence of those who died and suffering of those left behind, resisting the essentialist and reductive dimensions of resilience.

Moreover, conceptualising the 2010 disaster, beyond extremes, from experience, as recounted, fictionalised or imagined in a narrative form is an attempt to jointly account for the hinged chronologies of the event, their material manifestations, as well as the wider individual and collective cosmologies and ways of making sense and assigning meaning to an overwhelming and traumatic event. This includes engaging with Haiti's religious and spiritual landscape (directly referenced and interwoven in narratives in Part IV), which has long been defined by a dynamic, and at times violent, interaction between traditional Vodou practices and an increasing number of Christian denominations. These complexly intertwined,[68] in the Haitian context, religious and spiritual traditions provide sophisticated and composite conceptualisations of the relationship between personhood, emotions, embodiment and injustice, offering a highly specific understanding of mental illness and emotional distress in times of crises.

Vodou is, as Patrick Bellegarde-Smith and Claudine Michel assert, 'arguably the most maligned African-based religion in the Americas',[69] and its history has been subject to processes of silencing, erasure and denigration,[70] not least after the January disaster which was interpreted by a number of evangelical groups in terms of the rightful punishment of an idolatrous nation. Against such discriminatory portrayals, Vodou, continues to be Haiti's main spiritual language, offering, in Claudine Michel and Patrick Bellegarde-Smith's gloss, 'a spiritual discipline [as well as a] paradigmatic framework for a possible social, economic, and political transformation of Haiti'.[71] In the Vodou-inspired outlook, for example, 'the person is constituted of multiple parts and is situated at the nexus of social relationships among the lwa,[72] the ancestors, and the family'.[73] All of these are embodied relationships. The head is the most vital part of human consciousness.[74] Equally, *gwo bon anj* and *ti bon anj*, sometimes mistakenly translated as two souls, are rather 'two symbolic renditions that explain the physical (material) and metaphysical (psychic) life'.[75] In this and other culturally specific modes of embodiment,[76] the different psychosocial forces involved can 'cause imbalances between the body, the blood, and the emotions, which can destabilise personhood and the sense of self and can result in physical illness'.[77] These intricate notions of embodiment and personhood are the key source and inspiration for Lake's novel and the imaginaries of the future it drafts.

A lived and living cosmology, in its quotidian manifestations, Vodou is often joined by and seamlessly interwoven with other religious practices often those anchored in broadly-defined Christian theologies and denominations. This layered and embedded understanding of one's subjectivity directly shapes ideas surrounding individual and collective suffering and its meaning, for example as punishment or an opportunity to lead a better Christian life (Part IV), and the ways of alleviating it by turning to both God and the *loas*. Regardless of some obvious differences between them, in the wake of a disaster such as the 2010 earthquake in Haiti, religious beliefs, whether in the form of a defined theology or a braided set of frames and points of reference, can offer a sense of consolation and belonging to a wider community, in addition to that provided by one's circle of trusted associates, *moun pa*.[78] In effect, they contribute to a formulation of explanatory frameworks for the event by placing it as a part of a greater, if not immediately accessible, design. They create hinged and networked interpretative frames which allow the readers to place the disaster and assign its meaning within personal/collective cosmologies.

As a result, understandings of personal and collective impacts of traumatic events need to include these cultural and anthropological underpinnings, acknowledging that the varied spiritual frames of reference, and how they shape one's way of responding to a disaster and its violence, cannot be limited to a trajectory of rebound and could only be accurately understood in the context of Haiti's collective cosmology and history.[79] In this context of complex cosmologies, shifting religious belonging, 'strategic positioning'[80] and inclusive kinship within a web of long-term processes, whether environmental or historical, the immediate aftermath of the earthquake and the attempt to resume one's life emerge as *remnant dwelling*: that is, at once, a sense of living through the experience of violent fragmentation and an attempt to live through the fact of one's own survival and others' absence. This dual focus on the specific moment as well as the process, attends to the variety of 'sense-losing, improvising, sense-remaking, and renewing'[81] episodes that all are part of coming to terms with, and only sometimes overcoming, the experience of a disaster.

Together, this shift to *hinged and processual* notions of the 2010 disaster offered in my book builds on some of the *postcolonising* and *decolonising* reframings of disasters.[82] It is a move away from the limiting definitions and applied practices of resilience and recovery – terms that are intricately bound with and linked to practices of neoliberal development and imposed models of temporality – as a wished for return to a pre-disaster, vulnerable, normal. In contrast to such uses of the terms to frame a disaster as a 'development opportunity', emancipatory analytical approaches, to which this book contributes, would act as forms of anticipatory discourse and future-oriented critical practice, activating, in Anthony Carrigan's gloss, 'the legal and pecuniary connotations of recovery (as in recovering losses) *alongside* the need to "recover" occluded narratives and voices' [italics original].[83] In short, considering the 2010 disaster from experience and through ideas of *survenance/survivance* and dwelling is an essential prerequisite for envisaging post-disaster reconstruction and renewal – not as a return but rather re-conceptualised as transformation and improving over the long term.[84] In this sense, writing the self after the earthquake and post-earthquake rebuilding are better conceived in terms of remaking and *renewal* rather than recovery, i.e. an attempt to reconfigure, completely anew, one's here and now as embedded in the knowledge of the past and oriented towards possible futures. These asynchronous and hinged attempts reflect the non-linear, individualised and highly variable nature of an individual's and community's efforts to remake their lives, long after the tremors have ended, and slowly accept that the future is an unhealed one.

Notes

1   David Middleton and Steven D. Brown, *The Social Psychology of Experience Studies in Remembering and Forgetting* (London: SAGE, 2005), p. 81.
2   For an example of such discourses on Haiti's 'endemic weakness' and as continually weak, see: Marlye Gélin-Adams and David M. Malone, 'Haiti: A Case of Endemic Weakness', in *State Failure and State Weakness in a Time of Terror*, ed. by Robert I.

Rotberg (Cambridge, MA: World Peace Foundation; Washington, D.C: Brookings Institution Press, 2003), pp. 287–304. Another version of such formulation, that differentiates between weak and failed states, is Robert I. Rotberg's succinct diagnosis: 'Haiti seems condemned to remain weak, but without failing.' Robert I. Rotberg, 'Failed States, Collapsed States, Weak States: Causes and Indicators', in *State Failure and State Weakness in a Time of Terror*, pp. 1–26 (p. 19).

3 As Ingram and others make clear: 'Exposure is an external dimension of vulnerability (Füssel and Klein, 2006) that is typically a product of physical location and the characteristics of the surrounding built and natural environment (Pelling, 2003).' See Jane C. Ingram and others, 'Post-Disaster Recovery Dilemmas: Challenges in Balancing Short-Term and Long-Term Needs for Vulnerability Reduction', *Environmental Science & Policy*, 9 (2006), 607–13 (p. 610). This summary definition of exposure draws on the work of Füssel and Klein; also that of Pelling. See Hans-Martin Füssel and Richard J. T. Klein, 'Climate Change Vulnerability Assessments: An Evolution of Conceptual Thinking', *Climatic Change*, 75 (2006) 301–29; Mark Pelling, *The Vulnerability of Cities: Natural Disasters and Social Resilience* (London and Sterling, VA: Earthscan Publications, 2003).

4 Monique Terrier and others, 'Revision of the Geological Context of the Port-Au-Prince Metropolitan Area', 2577–87.

5 'The Enriquillo-Plantain Garden Fault has been silent since 1770 and the Septentrional Fault since 1842, when it leveled Cap Haitian.' Jean-Germain Gros, 'Anatomy of a Haitian Tragedy: When the Fury of Nature Meets the Debility of the State', *Journal of Black Studies*, 42 (2011), 131–57 (p. 148).

6 For an historical overview of Haiti's earthquakes see 'Haiti's History of Earthquakes', *Repeatingislands.com* https://repeatingislands.com/2010/01/15/haiti%E2%80%99s-history-of-earthquakes/ [accessed 21 February 2018]; and William H. Bakun, Claudia H. Flores, and Uri S. ten Brink, 'Significant Earthquakes on the Enriquillo Fault System', 18–30.

7 For a detailed analysis of the geological specifications of the 2010 Haitian earthquake, referred to in the study as the 2010 Léogâne earthquake, see: Gavin. P. Hayes and others, 'Complex Rupture During the 12 January 2010 Haiti Earthquake', *Nature Geoscience*, 3 (2010), 800–5 (p. 800).

8 Hayes and others, p. 800.

9 Hayes and others, p. 800.

10 Analysis by Hayes and others explains historic similarities between the events:

> The initial location and mechanism for this event suggested rupture on the Enriquillo-Plantain Garden fault zone (EPGF) [...]. The EPGF is the probable source of several large historic earthquakes in the region (National Oceanic and Atmospheric Administration (NOAA) National Geophysical Data Center (NGDC); www.ngdc.noaa.gov/hazard/hazards.shtml) including major events in November 1751 and June 1770. Both caused significant damage in Port-au-Prince; several hundred fatalities were directly attributed to the 1770 event (NOAA NGDC).
>
> Hayes and others, p. 803

11 Terrier and others, 'Revision of the Geological Context', p. 2586.

12 'Haiti', *The World Factbook* www.cia.gov/library/publications/the-world-factbook/geos/ha.html [accessed 13 September 2017]. In its 2015 estimate, Haitian Institute of Statistics and Information (L'Institut Haïtien de Statistique et d'Informatique (IHSI)) gives a slightly higher number of 10,911,819 inhabitants. L'Institut Haïtien de Statistique et d'Informatique, 'Statistiques Démographiques & Sociales', www.ihsi.ht/accueil_presentation_general.htm [accessed 16 February 2018].

13 Myrtho Joseph, Fahui Wang, and Lei Wang, 'GIS-based assessment of urban environmental quality in Port-au-Prince, Haiti', *Habitat International*, 41(2014), 33–40 (p. 33).

14   Urban areas also are especially susceptible to the effects of natural disasters. With mounting rural–urban migration, poor urban neighbourhoods spread to ever-more marginal land, which means that when an earthquake, hurricane, or mudslide occurs, they are affected disproportionately. In addition, of course, the potential for epidemics is much higher in urban areas because of the concentration of people.

   Elizabeth Ferris and Sara Ferro-Ribeiro, 'Protecting People in Cities: the Disturbing Case of Haiti', *Disasters*, 36 (2012), 44–63 (p. 48)

15   Harley F. Etienne, 'Urban Planning and the Rebuilding of Port-au-Prince', in *The Idea of Haiti*, pp. 165–81 (p. 168). The whole chapter is a critical and lucid discussion of the needs and challenges to urban planning in post-earthquake Haiti.

16   Guitele J. Rahill, N. Emel Ganapati, J. Calixte Clérismé, and Anuradha Mukherji, 'Shelter recovery in urban Haiti after the earthquake: the dual role of social capital', *Disasters*, 38 (S1) (2014), S73–S93 (p. S73).

17   Rahill and others, p. S73.

18   Frank Graziano, *Undocumented Dominican Migration* (Austin: University of Texas Press, 2013), p. 4.

19   Graziano, p. 5.

20   Graziano, p. 2.

21   Clemens Höges, 'Preparing for the Next Earthquake: Haiti Debates Moving Its Capital', *Spiegel*, www.spiegel.de/international/world/preparing-for-the-next-earthquake-haiti-debates-moving-its-capital-a-675299–3.html, 2 February 2010 [accessed 14 December 2017]

22   Terrier and others, 'Revision of the Geological Context of the Port-Au-Prince Metropolitan Area', p. 2586.

23   'Haiti Earthquake PDNA: Assessment of damage, losses, general and sectoral needs' (2010) https://siteresources.worldbank.org/INTLAC/Resources/PDNA_Haiti-2010_Working_Document_EN.pdf, p. 24 [accessed 12 October 2017]. This Post-Disaster Needs Assessment (PDNA) report was prepared by a joint team composed of representatives of the government and members of the international community, under the direction of the government of the Republic of Haiti.

24   Richard Pallardy, 'Haiti earthquake of 2010', *Encyclopædia Britannica*, 28 June 2017, www.britannica.com/event/Haiti-earthquake-of-2010 [accessed 12 October 2017].

25   'Frequently Asked Question – Haiti Earthquake – 2010' (St. Augustine, Trinidad and Tobago: UWI, Seismic Research Centre) http://uwiseismic.com/Downloads/2010_01_HaitiEQ_FAQ2c.pdf [accessed 12 October 2017].

26   Jeff Malpas, *Place and Experience: A Philosophical Topography* (Cambridge: Cambridge University Press, 1999), p. 35.

27   Homel, p. 82.

28   Erica Caple James, 'Ruptures, Rights, and Repair: The Political Economy of Trauma in Haiti', *Social Science & Medicine*, 70 (2010), 106–13 (107). For an extended discussion of *ensekirite* see James's earlier piece: 'Haunting Ghosts: Madness, Gender, and Ensekirite in Haiti in the Democratic Era', in *Postcolonial Disorders*, ed. by Mary-Jo Delvecchio Good, Sandra Teresa Hyde, Sarah Pinto, and Byron J. Good (Berkeley: University of California Press, 2008), pp. 132–56.

29   See Alessandra Benedicty, 'Aesthetics of "Ex-centricity" and Considerations of "Poverty"', *Small Axe*, 16 (2012), 166. As Alessandra Benedicty-Kokken makes clear:

   What is clear then is that: primary measurement for poverty is linked up to nationhood; that the indices for determining 'poverty' are primarily economic, and 'based on scientific evidence'; and that the vocabulary used for studying poverty is, at least in the current context, unavoidably degrading, based on a civilizing mission, dividing nations into those that are 'developed' and those that aren't.

   Alessandra Benedicty-Kokken, *Spirit Possession in French, Haitian, and Vodou Thought: An Intellectual History* (Lanham, MD: Lexington Books, 2015), p. 87

30 'UN Radio: Poverty is behind Haiti's vulnerability to natural disasters' [interview], *Preventionweb.net* www.preventionweb.net/go/12306 [accessed 15 September 2017].

31 Alyssa Goldstein Sepinwall, 'Introduction', in *Haitian History: New Perspectives*, ed. by Alyssa Goldstein Sepinwall (New York, NY: Routledge, 2013), pp. 1–13 (p. 7).

32 Alyssa Goldstein Sepinwall in the Introduction to *Haitian History: New Perspectives* points to the growing acknowledgement of the importance of the Haitian Revolution (especially in the English-speaking academic world) which is not, however, matched with an equal scholarly interest in the later periods of Haiti's history. For her full discussion of the Haitian Revolution and its historical and cultural representations see Alyssa Goldstein Sepinwall, 'Representations of the Haitian Revolution in French literature', in *The World of the Haitian Revolution*, ed. by David Patrick Geggus and Norman Fiering (Bloomington, IN: Indiana University Press, 2009), pp. 339–52. For a recent study of artistic representations of the Haitian Revolution and the political visions they evoked, see Philip Kaisary, *The Haitian Revolution in the Literary Imagination: Radical Horizons, Conservative Constraints* (Charlottesville and London: University of Virginia Press, 2014); as well as Victor M. Figueroa, *Prophetic Visions of the Past: Pan-Caribbean Representations of the Haitian Revolution* (Columbus, OH: Ohio State University Press, 2015).

33 Nick Nesbitt, 'Alter-Rights: Haiti and the Singularization of Universal Human Rights, 1804–2004', *International Journal of Francophone Studies*, 12 (2009), 93–108 (p. 93).

34 Kate Ramsey in *The Spirits and the Law: Vodou and Power in Haiti* (Chicago and London: University of Chicago Press, 2011) extensively discusses the scholarly controversies and uncertainties around the Bwa Kayiman ceremony and the myriad of ways the event, regardless of its exact details, functions as a reference point as well as an integral part of collective memory, especially for those community living the region. As she writes:

> the scholarly debate over Bwa Kayiam is illuminated and complicated by those with a lengthy ancestry in the region who remember and narrate this history, and understand its symbolism and legacies in relation to specific physical sites and through particular spiritual practices.
>
> Ramsey, p. 43

At the same time, it is important to note how the ceremony is taken up in negative and further demonising portrayals of Haiti and Vodou, being presented – a trope taken up by Robertson and other media commentators – as just another example, or the *ur*-example, of Vodou's diabolic essence and, by extension, Haiti's excessive irrational beginnings as a nation.

35 Peter Hallward, *Damming the Flood: Haiti and the Politics of Containment* (London, New York, NY: Verso, 2010), p. 74.

36 Alyssa Goldstein Sepinwall, 'From the Occupation to the Earthquake: Haiti in the Twentieth and Twenty-First Centuries', in *Haitian History: New Perspectives*, pp. 215–41 (p. 223).

37 Beckett, p. 41.

38 Jean-Bertrand Aristide, *Eyes of the Heart: Seeking a Path for the Poor in the Age of Globalization* (Monroe, ME: Common Courage Press, 2000), pp. 16–17; in Hallward, p. 55.

39 See, for example, Human Rights Watch 2001 Report or Robert Fatton Jr's, *Haiti's Predatory Republic: the Unending Transition to Democracy* which traces back the roots of the electoral crisis and Haiti's political impasse. Robert Fatton Jr., *Haiti's Predatory Republic: The Unending Transition to Democracy* (Boulder, CO: Lynne Rienner Publishers, 2002); Human Rights Watch, *2001 Report*, www.hrw.org/legacy/wr2k1/americas/haiti.html [accessed 19 January 2016].

40 An issue of *Small Axe* 13 (2009) was a forum for an extensive debate between Alex Dupuy, Lyonel Trouillot, Peter Hallward, Valerie Kaussen and Nick Nesbitt on the

question of violence, and its scale, during Aristide' s presidency. Hallward's assessment of it in *Damning the Flood* is in stark contrast to Dupuy's analysis in Alex Dupuy, *The Prophet and Power: Jean-Bertrand Aristide, the International Community, and Haiti* (New York, NY: Rowman & Littlefield Publisher, 2007).

41 Created in 1959 by François Duvalier, this paramilitary squadron of loyal troops effectively replaced Haiti's armed forces, 'perpetua[ted] popular animosity toward an ambiguously defined mulatto elite and orchestra[ted] a reign of terror that would keep him [Papa Doc] in power'. David Sheinin, 'The Caribbean and the Cold War: Between Reform and Revolution', in *The Caribbean: A History of the Region and Its Peoples*, ed. by Stephan Palmié and Francisco A. Scarano (Chicago, IL. and London: University of Chicago Press, 2011), pp. 491–505 (pp. 498–9). Following the murder of his colleague Gaston Raymond, Laferrière decided to move to Montréal in 1978. 'French-Canadian Writers: Dany Laferrière', *Athabasca University*, http://canadian-writers. athabascau.ca/french/writers/dlaferriere/dlaferriere.php [accessed 10 February 2018].

42 It is estimated that the UN-caused cholera outbreak has claimed over 11,000 lives so far and afflicted another 932,774. Just Foreign Policy, *Haiti Cholera Counter* www. justforeignpolicy.org/haiti-cholera-counter [accessed 16 February 2018]. The Boston-based Institute for Justice and Democracy in Haiti (IJDH) has been leading the Cholera Accountability project and

> works with Haitian grassroots groups and international advocates in a broad-based campaign to force the UN to take action to stop the cholera's killing. Diarrheal disease and lack of safe water are the worst public health problems in Haiti. Our case demands the installation of water and sanitation infrastructure that will control the epidemic and save more than 5,000 lives each year.
>
> IJDH, 'Cholera Accountability', *IJDH.org*, www.ijdh.org/advocacies/our-work/
> cholera-advocacy/[accessed 16 February 2018]

For a comprehensive overview of the case against the UN as well as a list of media coverage, reports and petitions, see: IJDH, 'Cholera Resources', *IJDH.org*, www.ijdh. org/cholera/cholera-resources/ [accessed 16 February 2018].

43 Alyssa Goldstein Sepinwall in the Introduction to *Haitian History: New Perspectives* points to the growing acknowledgement of the importance of the Haitian Revolution (especially in the English-speaking academic world) which is not, however, matched with an equal scholarly interest in the later periods of Haiti's history. For her full discussion of the Haitian Revolution and its historical and cultural representations see Alyssa Goldstein Sepinwall, 'Representations of the Haitian Revolution in French literature', in *The World of the Haitian Revolution*, ed. by David Patrick Geggus and Norman Fiering (Bloomington, IN: Indiana University Press, 2009), pp. 339–52.

44 Sarah Pulliam Bailey, 'Pat Robertson: Haiti "Cursed" Since Pact with the Devil', *Christianity Today*, 13 January 2010 www.christianitytoday.com/gleanings/2010/ january/pat-robertson-haiti-cursed-since-pact-with-devil.html?paging=off [accessed 11 February 2015]. Another example of an objectifying approach, which uncritically mirrors colonial and neoliberal discourses, is David Brook's article 'The Underlying Tragedy'. In it, Brooks emphasises the inefficacy of international aid and its lack of real impact, asserting that

> Haiti, like most of the world's poorest nations, suffers from a complex web of progress-resistant cultural influences. There is the influence of the voodoo religion which spreads the message that life is capricious and planning futile. There are high levels of social mistrust. Responsibility is often not internalized. Child-rearing practices often involve neglect in the early years and harsh retribution when kids hit 9 or 10.
>
> David Brooks, 'The Underlying Tragedy', *New York Times*, 15 January 2010
> www.nytimes.com/2010/01/15/opinion/15brooks.html?_r=0
> [accessed 25 February 2014]

45 Nadège T. Clitandre, 'Haitian Exceptionalism in the Caribbean and the Project of Rebuilding Haiti', *Journal of Haitian Studies*, 17 (2011), 146–53 (p. 146).

46 Molefi Kete Asante analyses earlier representations of Haitian history as well as the 2010 earthquake within 'a tripartite framework based on the *class narrative*, the *religious narrative*, and the *cultural narrative*' (p. 277; emphasis original). Stressing that the earthquake is 'a natural geological action that has nothing to do with Haitian capability, character, or potential' (p. 286), he underlines the need to confront, in the aftermath of the event, 'the organizational, institutional, social, and economic issues that may not have been radically changed in the natural disaster' (p. 286). Molefi Kete Asante, 'Haiti: Three Analytical Narratives of Crisis and Recovery', *Journal of Black Studies*, 42 (2011) 76–87.

47 An entire special issue of *Journal of Black Studies*, 42 (2011), entitled 'The Haiti Earthquake of 2010: The Politics of a Natural Disaster', was published in the wake of the event as a response to the speculation and misinformation that would flood the internet, and the media more generally, immediately after the disaster. As Garvey Lundy, the editor of the issue, makes clear:

> Most of the information presented was ahistorical and bordered on gross racial caricature. What little context was provided was compressed into 30-second sound bites that did an injustice to the complexity of Haitian society and history. Ironically, many of the contributors to this issue were called upon by media outlets to provide insights into the earthquake, but given the structure of the news, they were allowed, in most cases, just a few minutes to expound on a situation that was worthy of a book.
>
> Garvey Lundy, 'The Haiti Earthquake of 2010: The Politics of a Natural Disaster', *Journal of Black Studies*, 42 (2011), 127–30 (p. 128)

48 Ilan Kelman and others, 'Learning from the history of disaster vulnerability and resilience research and practice for climate change', *Natural Hazards*, 82 (2016) S129–S143 (p. S131).

49 David E. Alexander, 'Resilience and disaster risk reduction: an etymological journey', *Natural Hazards and Earth System Sciences*, 13 (2013), 2707–16; Judite Blanc and others, 'Religious Beliefs, PTSD, Depression and Resilience in Survivors of the 2010 Haiti Earthquake', *Journal of Affective Disorders*, 190 (2016) 697–703.

50 Siambabala Bernard Manyena, 'The concept of resilience revisited', *Disasters*, 30 (2006), 433–50; Ben Wisner and Ilan Kelman, 'Community Resilience to Disasters', *International Encyclopedia of the Social & Behavioral Sciences Online* (Oxford: Elsevier, 2015), pp. 354–60 https://doi.org/10.1016/B978-0-08-097086-8.28019-7 [accessed 1 February 2018]; J. C. Gaillard, 'Resilience of traditional societies in facing natural hazards', *Disaster Prevention and Management*, 16 (2007), 522–44.

51 Crawford Stanley Holling, 'Resilience and Stability of Ecological Systems', *Annual Review of Ecology and Systematics*, 4 (1973), 1–23.

52 American Psychological Association, 'The road to resilience' (Washington, DC: American Psychological Association, 2014) www.apa.org/helpcenter/road-resilience. aspx [accessed 23 October 2017].

53 Carl Folke and others, 'Resilience Thinking: Integrating Resilience, Adaptability and Transformability', *Ecology and Society*, 15 (2010) www.ecologyandsociety.org/vol. 15/iss4/art20/ [accessed 16 September 2017].

54 Steven M. Southwick and others, 'Resilience Definitions, Theory, and Challenges: Interdisciplinary Perspectives', *European Journal of Psychotraumatology*, 5 (2014) 10.3402/ejpt.v5.25338 [accessed 25 September 2017].

55 Guitele Rahill and others, 'In Their Own Words: Resilience among Haitian Survivors of the 2010 Earthquake', *Journal of Health Care for the Poor and Underserved*, 27 (2016) 580–603, Box 1 and 2 (pp. 586–7).

56 Luckhurst, *The Trauma Question*, p. 15.

57 Ilan Kelman, Jessica Mercer, and Jennifer West, 'Combining Different Knowledges: Community-based Climate Change Adaptation in Small Island Developing States', *Participatory Learning and Action Notes*, 60 (41–53); Ilan Kelman, James Lewis, JC Gaillard, and Jessica Mercer, 'Participatory Action Research for Dealing with Disasters on Islands', *Island Studies Journal*, 6 (2011), 59–86; Ilan Kelman, 'How resilient is resilience?', Urban Resilience Research Network, 20 April 2016, www. urbanresilienceresearch.net/2016/04/20/how-resilient-is-resilience/ [accessed 2 February 2018].

58 For an analysis of how and to what ends resilience is defined and employed, see: Julian Reid, 'The Disastrous and Politically Debased Subject of Resilience', *Development Dialogue: The End of the Development-Security Nexus? The Rise of Global Disaster Management*, 58 (2012), 67–79; Siambabala Bernard Manyena, 'The concept of resilience revisited', *Disasters*, 30 (2006), 433–50; David E. Alexander, 'Resilience and disaster risk reduction: an etymological journey', *Natural Hazards and Earth System Sciences*, 13 (2013) 2707–16.

59 Rahill and others, 'In Their Own Words', p. 600.

60 Kaiama L. Glover, 'New Narratives of Haiti; or, How to Empathize with a Zombie', *Small Axe*, 16 (2012), 199–207 (p. 200).

61 Glover, p. 201.

62 Glover, p. 201.

63 Glover, p. 200.

64 Nicolas Guerda and others, 'Weathering the Storms Like Bamboo: The Strengths of Haitians in Coping with Natural Disasters', in *Mass Trauma and Emotional Healing Around the World: Rituals and Practices for Resilience and Meaning-Making*, ed. by Ani Kalayjian and Dominique Eugene (Santa Barbara, CA: Praeger, 2010), pp. 93–106 (p. 97).

65 Edwidge Danticat, contesting the ways in which the trope of Haitians' resilience was used after the earthquake to justify the international community's neglect, draws attention to the long-term suffering the affected population is subject to:

> After three post-earthquake visits to Haiti, I began to ask myself if this much-admired resilience would not in the end hurt the affected Haitians. It would not be an active hurt, like the pounding rain and menacing winds from the hurricane season, the brutal rapes of women and girls in many of the camps, or the deaths from cholera. Instead, it would be a passive hurt, as in a lack of urgency or neglect. 'If being resilient means that we're able to suffer much more than other people, it's really not a compliment,' a young woman at the large Champs de Mars camp in downtown Port-au-Prince told me.
>
> Edwidge Danticat, 'Lòt Bò Dlo, The Other Side of the Water', in *Haiti After the Earthquake*, ed. by Paul Farmer (New York, NY: Public Affairs, 2011), pp. 249–59 (p. 257)

66 Pascale Monnin, Interview with the artist, 17 July 2017.

67 Kas Saghafi, 'Dying Alive', *Mosaic: a journal for the interdisciplinary study of literature*, 48 (2015), 15–26 (p. 22) [italics original].

68 For a discussion of the many paradoxes and likely symbiosis between Vodou and Roman Catholicism in Haiti, see, for example: Leslie G. Desmangles, *The Faces of the Gods: Vodou and Roman Catholicism in Haiti* (Chapel Hill, NC: University of North Carolina Press, 1992); for an extended analysis of the relationship between Catholicism and human rights, see: Terry Rey, 'Catholicism and Human Rights in Haiti: Past, Present, and Future', *Religion & Human Rights*, 1 (2006), 229–48.

69 Patrick Bellegarde-Smith and Claudine Michel, 'Introduction', in *Haitian Vodou: Spirit, Myth, and Reality* ed. by Patrick Bellegarde-Smith and Claudine Michel (Bloomington: Indiana University Press, 2006), xvii–xxvii (p. xix).

70 Laurent Dubois, in an analysis long preceding the earthquake, examines the ways in which Vodou, and the scholarship on it, have developed in response to Haiti's histories of oppression, displacement and contemporary migration. Experiences of slavery and plantation economy were foundational for the development of Haitian Vodou. Other key moments that have shaped it include

> the re-emergence of the foreign priesthood in Haiti, the development of the urban centers, the U.S. occupation of 1915–34, which incited new revolts and new cultural developments, the Duvalier dictatorship and the dechoukaj that struck many ougans and manbos (priestesses) in its wake, not [p. 95] to mention the creation of a massive Haitian diaspora which itself is auguring a new set of transformations in the religion.
> Laurent Dubois, 'Vodou and History', *Comparative Studies in Society and History*, 43 (2001), 92–100 (pp. 94–5)

For a masterfully written and unmatched analysis of anti-Vodou rhetoric, practices and legislation, see: Kate Ramsey, *The Spirits and the Law: Vodou Power in Haiti* (2011).

71 Bellegarde-Smith and Michel, *Haitian Vodou*, 'Introduction', xxi.

72 Lwa (*loa*) are spirits in the Haitian Vodou tradition. They are intermediaries between God the Creator and the humanity. They are supernatural beings that are thought to be present in all realms of nature and create a web of linkages between human activities and various aspects of human life. They can enter the human body, interfere in the affairs of the living, and take control of an individual's life from birth to death. Also, they explain the origins of the world, providing a way of classifying life in society that takes on meaning through the agency of the *lwa*. *Lwa* are believed to come from the mythical Guinea, and the different categories of *lwa* reflect the various African ethnic groups. *Lwa* are regrouped into families called nations (*nanchon*) and each has distinctive ceremonies and rituals. The three important rituals are Rada, Kongo and Petro. *Lwa* carry different weight and importance in each of the pantheons corresponding to the three main rituals. Laënnec Hurbon, *Voodoo Search for the Spirit* (New York, NY. and London: Harry N. Abrams, 1995), pp. 65–72.

73 Erica Caple James, 'Haiti, Insecurity, and the Politics of Asylum', *Medical Anthropology Quarterly*, 25 (2011), 357–76 (p. 360).

74 Guérin C. Montilus, 'Vodun and Social Transformation in the African Diasporic Experience: The Concept of Personhood in Haitin Vodun Religion', in *Haitian Vodou*, pp. 1–7 (p. 2).

75 Montilus, p. 2. This is also key in understanding the period of possession by the lwa where *gwo bon anj* (the psychic principle) leaves the body and comes back after the trance. 'However, the *ti bon anj* remains present, otherwise death would take place, for the body would have lost the vital principle which keeps it alive.' Montilus, p. 4.

76 Erika Bourguignon, 'Belief and Behavior in Haitian Folk Healing', in *Mental Health Services: The Cross-Cultural Context*, ed. by Paul B. Pedersen, Norman Sartorius, and Anthony J. Marsella (Beverly Hills, CA: Sage 1984), pp. 243–66 (p. 261).

77 James, 'Haiti, Insecurity, and the Politics of Asylum', p. 360.

78 Guitele J. Rahill and others, 'Shelter recovery in urban Haiti after the earthquake', p. S75.

79 Kari A. O'Grady and others, 'Earthquake in Haiti: Relationship with the Sacred in Times of Trauma', *Journal of Psychology & Theology*, 40 (2012), 289–301.

80 Karen Richman, 'Religion at the Epicenter: Agency and Affiliation in Léogâne After the Earthquake', *Studies in Religion/Sciences Religieuses*, 41 (2012), 148–65 (p. 152).

81 James Douglas Orton and Kari A. O'Grady, 'Cosmology Episodes: A Reconceptualization', *Journal of Management, Spirituality & Religion*, 13 (2016), 226–45 (p. 226).

82 Anthony Carrigan, 'Towards a Postcolonial Disaster Studies', pp. 117–40.

83 Carrigan, p. 131.

84 Ben Wisner and Ilan Kelman, 'Community Resilience to Disasters', pp. 359, 360; James Lewis, 'Natural Hazard Reduction', *Environment*, 30 (1988), 3–4 (p. 4).

3 *Rasanblaj*

Scholarship of care

Weaving together distinct insights on the earthquake, *Disasters, Vulnerability, and Narratives: Writing Haiti's Futures* is a *rasanblaj*, a Haitian Kreyòl term that encompasses, in Gina Athena Ulysse's poignant and rigorous meditation of it, '*[an] assembly, compilation, enlisting, regrouping (of ideas, things people, spirits. For example, fè yon rasanblaj, do a gathering, a ceremony, a protest)*' [italics original].[1] This book, then, is such a gathering of voices and a bringing together rather than opposing, of mutually enriching perspectives on the January tremors, that aims to commemorate the recent losses and past hurt and to imagine empowered, non-catastrophic futures. Across the subsequent sections this book hopes to then take up and respond to Ulysse's call for this practice of *rasanblaj* that would

> Emerg[e]
> Out of
>
>   epistemic conflicts
>   &embodied nightmares
>
> embodied conflicts
> &epistemic nightmares.[2]

Bringing together contrasting epistemologies – ways of knowing, explaining and acting upon the experience of the disaster and its past, present and future – this book takes selected post-earthquake narratives as point of entry into exploration of categories of time, place and self and the cognate notions of disaster time/ duration, reconstruction/ruination and recovery/return. These are rooted in specific understandings of when the disaster happened and how long it lasted (time), where did it happen and what imprint it left on people and their surroundings (space), and how the disaster's impact on individuals (self) and the community can be mitigated in the years following upon it. This book then, in three movements, traces these conflicts, tensions and contrasts across the varied texts and voices, offering a rigorously polyphonic account of the past and present of the January tremors and the future that emerges in their aftermath.

'Part II: Disaster Time' – focusing on time, temporality and the history of the disaster – is a comparative study of Martin's *Le Tremblement*: *Haïti, le 12 janvier 2010* (2010) and Saint-Éloi's *Haïti kenbe la! 35 secondes et mon pays à reconstruire* (2010). These vivid survivors' accounts, despite significant formal differences, already in the title point to their key preoccupation with disaster time and both reveal the difficulties of locating temporal reference points within which to narrate the traumatic experience of the earthquake. Both view the event as the marker of a new history – but one that is defined by a haunting presence of the disaster. The two texts, by demonstrating their awareness of literary form and its limits and by questioning the conventional methods of temporal arrangement they use, reflect on their own inability to control a traumatising experience within the confines of objective duration and time. What is revealed, the chapter suggests, is the *untimely*, after Ross Chambers, character of the earthquake and of its narrative representations.[3] Consequently, these narratives directly challenge standard categories of linear narrative composition and the notion of disaster as a time-bound and disconnected event, indicating the difficulty of distinguishing between the event and its aftermath. This processual view is directly relevant for discussion of other regions which also suffer from compound vulnerabilities and greater exposure to environmental hazards.

'Part III: Disaster Space' – concentrating on place and landscape, their pre- and post-disaster reshapings – directly builds on this notion of untimeliness in order to examine the different ways in which the past, present and future of local island landscapes are reconfigured and envisaged. It compares Dany Laferrière's collection of sketches focused on the everyday vulnerabilities and defiant lives among ruins, *Tout bouge autour de moi* (2011), and Sandra Marquez Stathis's *Rubble: The Search for a Haitian Boy* (2012), an account of her journey to Haiti in the aftermath of the event and a pledge of commitment to the future of the titular Haitian boy. Both texts cross Haiti's capital, navigating through the debris and pausing at the same yet contrastingly experienced sites, and explore the relationship between a radically altered landscape and personal and cultural memory of it, and between the broader sweep of the country's history. The analysis places an emphasis on the *temporal* dimensions of space and landscape – the histories it embodies, the tensions it exposes, and, more crucially, the possible futures it permits itself to imagine – demonstrating the oft asynchronous temporalities of remaking, reconstruction and rebuilding. In the process of forging a narrative form and language that would at least index the 'before' and 'after' of the disaster, they explore the central issues of post-disaster literary endeavours, namely: how to speak of and share the experience of the highly particular yet context-specific event and its unparalleled, life-redefining impact without reproducing and rehearsing the very images and discourse on Haiti's uniqueness, seen only in extreme terms.

'Part IV: Disaster Selves' – centring on notions of subjectivity, self and recovery – examines the unsettling effect of the tremors, of the experience of being entrapped in ruins, and of the post-earthquake landscapes the quake left behind on foundational categories of self and other and considers possible forms personal and collective recovery might take in Dan Woolley's *Unshaken: Rising from the Ruins of Haiti's Hotel Montana* (2011), Nick Lake's *In Darkness*

(2012) and Laura Rose Wagner's *Hold Tight, Don't Let Go* (2015). The chapter brings together texts where twinning is both a formal organising principle and a key thematic concern, directly linked to ideas of starting again, countering and only sometimes overcoming, the sense of subjective emptiness and wholeness explored across the three works. The chapter starts off by analysing some of the ways in which the texts' respective narrators present the earthquake as a trigger for subjective transformation, whether this change is understood through a specifically religious framework (Woolley), is portrayed as a general metamorphosis, that is the ontological process of becoming someone else (Lake) or entails the acceptance of double loss and departure (Wagner). The chapter pauses also over the visions of collective renewal, probing in particular the significance of parallel narrative strands in Lake's text and the analogy between his narrator's life and that of Toussaint Louverture in the context of the 1791–1804 Haitian Revolution. The analysis asks whether the narrative frames the earthquake as offering a moment of redefinition of the Haitian state – an opportunity, like the Revolution, to claim political agency – or whether it is more interested in drawing a contrast between the post-earthquake dynamics of objectification and disempowerment and the subjective and collective affirmation of 1804. The chapter then turns to Wagner's novel which follows the story of two cousins who remain inseparable even after the earthquake until one of them, Nadine, leaves to join her father in the United States. Here, the fictional account of two girls' parallel lives allows the readers to further contemplate ideas for collective and individual recovery for Haiti and its *Dyaspora* (diaspora in Haitian Kreyòl) in the aftermath of the January disaster. Across the three texts, recovery, from an identifiable post-disaster state, becomes *recovering*. It is an ongoing and open-ended process, an in-between, that escapes the easy binaries of a restored past fullness or, its opposite, the sense of a permanent and unalterable break. Following on from the considerations of past and present, central to Parts II and III, this chapter then has a clear future-oriented focus and points to imaginaries of lives and times beyond the ruins – questions taken up again in *Rasanblaj*.

In '*Rasanblaj*: A Future Reassembled', the closing chapter of the book, the discussion returns to the more overarching lines of inquiry in the context of the ever more apparent inability of humanitarian aid to provide a sustainable vision for Haiti's post-earthquake recovery and the region's vulnerability to environmental hazards. Moreover, the chapter considers the wider methodological implications the insights gained via a close-reading of selected narratives might have for our understanding of disasters and practices of post-disaster reconstruction and recovery in Haiti. This includes an emphasis on the significance of post-earthquake literary production, although this is bound by its own limitations, and on its potential to offer insights into the individual experience of the event as well as the embedded character of the 2010 disaster that, in some forms, might not be over yet. Here, I return to the Haitian Kreyòl term *rasanblaj* (assembly, regrouping but also in *fè yon rasanblaj*, do a gathering, a ceremony, a protest), a guiding principle for the book's methodology, to reaffirm the importance of approaching disasters as slow processes with distinct histories, occurring in highly specific contexts of constructed and sustained vulnerability. Consequently, post-disaster reconstruction and recovery must address the longer histories of each event and work towards transforming the conditions of vulnerability of each particular place as it to struggles to combat the vulnerability imposed upon it by colonial past, by repeated ecological exploitation and by the pressures of the current neoliberal global economy.

In this context, the 2010 earthquake and the literature it has inspired have a crucial role to play in advancing new narratives of Haiti and new approaches to disasters, vulnerability, reconstruction and recovery. These differentiated framings and varied accounts enhance our critical understanding of these complex processes, partake in debates that do not have an easily identifiable solution, forging, in effect, a collaborative dialogue between researchers, activists and practitioners who, regardless of their disciplinary standpoints, share a scholarly-ethical commitment to a necessarily slow yet no less urgent work of reconstitution. Forging such 'sustainable scholarly collaborations between Haiti, the *dyaspora* and allies'[4] is equally at the centre of Haitian Studies, a scholarly community that, in the words of Claudine Michel, 'is at once a discipline and *interdiscipline*' [italics original].[5] It is an interspace with its 'geographic reach extend[ing] to the diverse expressions of *dyasporas* throughout the world and its spheres of influence both such physical landscapes as the Americas, Europe and the Caribbean, and ideological spaces, like the transatlantic and the Afro-Atlantic worlds'.[6] The series of close-readings and analyses of the earthquake offered in this book contribute to this assemblage, connecting these points of scholarly concern and formulations of care. After 12 January 2010, 'Haiti will never be the same. And must not be the same.'[7] Likewise, unidimensional definitions of and discipline-bound approaches to disaster, reconstruction and recovery fail to convey the complexity of the hinged disaster experience, its past, present and the unhealed futures charted in its aftermath. The 2010 Haiti earthquake is not the first nor is it the last disaster to strike the island. More recently, the aftermath of the 2016 Hurricane Matthew or 2017 Hurricane Irma exposed the compound vulnerability of Haiti's southern region and the wider Caribbean. Yet, refusing to give into the portrayals of Haiti's history as an ongoing disaster, this book tunes into the contradictions and probes the knots of a past event, that is not past, in order to envisage modes of empowered reconstruction and a future beyond repetition.

In effect, the series of close-readings offered here are in themselves critical acts of care: an attempt to formulate

> a kind of affectively and ethically engaged scholarship; one that also works to position [one's] writing, speaking and teaching – however modest their impacts – as *practical* acts of care that can draw others into a sense of curiosity and concern for our changing world.[8]

To this end, the book hopes to be a site of encounter, *un lieu de rencontre*, between the different disciplinary approaches that allow us to see the object, the

2010 disaster, at the crossroads of distinct perspectives. It privileges a moment of reflective pause and the acceptance of the event's complexity and lack of an easy solution, forming a space of scholarly care for and in a convalescent world. Finally, the critical conversations which this book offers are moments of reflective dialogue, attentive listening and rigorous and careful analysis. They are moments of recalling the losses and calling forth a vision, rooted in care, of open-ended futures, full of resolve.

Notes

1 Gina Athena Ulysse, 'Introduction', *e-misférica: Caribbean Rasanblaj* http://hemisphericinstitute.org/hemi/en/emisferica-121-caribbean-rasanblaj/e-121-introduction [accessed 18 December 2017].
2 Ulysse, 'Introduction'.
3 Ross Chambers, *Untimely Interventions: AIDS Writing, Testimonial, and the Rhetoric of Haunting* (Ann Arbor, MI: University of Michigan Press, 2004).
4 Claudine Michel, '*Kalfou Danje*: Situating Haitian Studies and My Own Journey within it', p. 196.
5 Michel, p. 197.
6 Michel, p. 197.
7 Gina Athena Ulysse, *Why Haiti Needs New Narratives*, p. 7.
8 Thom van Dooren, 'Care' *Environmental Humanities*, 5 (2014), 291–4 (p. 293) https://doi.org/10.1215/22011919–3615541 [accessed 10 October 2016]; see also Thom van Dooren and Deborah Bird Rose, 'Lively Ethography: Storying Animist Worlds', *Environmental Humanities*, 8 (2016): 77–94. doi: https://doi.org/10.1215/22011919–3527731 [accessed 12 February 2018].

Part II
Disaster time

4 Halting the tremors

Defined both as a moment of rupture, a marker of new history and as another example of continuing historical violence, the 2010 Haitian earthquake is often encapsulated in the symbolic date of its occurrence: 12 January 2010. After the first tremors, in the six weeks following the disaster there were approximately 59 aftershocks of magnitude Mw4.5 or greater; 16 of these were magnitude Mw5.0 or greater. The most powerful aftershock, registered at magnitude Mw6.0, occurred seven minutes after the main tremors on 12 January.[1] The sense of measured precision, which is conveyed by these figures, contrasts with subjective perceptions of the event and its duration. The disjunction between chronology and temporally shifting experiences of the catastrophe is at the centre of two survivor narratives written immediately after the earthquake: *Le Tremblement: Haïti, le 12 janvier 2010* (*Tremors: Haiti, January 12th, 2010*, 2010), by Lionel-Édouard Martin, and *Haïti, kenbe la! 35 secondes et mon pays à reconstruire* (*Haiti, hold on tight! 35 Seconds and my Country to Rebuild*, 2010), by Rodney Saint-Éloi. Both texts discussed in the chapter, on a formal and thematic level, explore questions of time and temporality as they try to present an overwhelming event that escapes easy temporal categorisation. The two post-earthquake narratives play with notions of narrative time and temporality – and related ideas of untimeliness, recalling the past, and temporal suspension – in their attempts to account for the overpowering experience of the disaster that seems to continue long after the initial tremors have ended.[2] In so doing, these narratives, while trying to convey the immediacy of the incident through their form, question the sense of the disaster as an event, equally recognising that memories of what happened are filtered through several layers of mediation.

It is above all the event's *untimely* character that defines Saint-Éloi's and Martin's accounts of the earthquake, and which informs the ethical interventions that both authors are trying – not entirely successfully – to make. As post-earthquake narratives of the country's recent and more distant past that are critically attuned to their own limitations, they effect the work of double recall as *remembering* and *reclaiming*, calling into question and transforming the ways in which Haiti's history has been discursively framed. Instead, they suggest collectively empowering ways of narrativising the island's past and imagining the country's future. This commitment to forging new forms of writing about the disaster is rooted in an acknowledgement that, like other untimely events, the earthquake cannot and should not be forgotten.[3] Immersed in the open-ended 'after' of the event, these works realise their duty to witness to the scale of the catastrophe for the sake of all those lost under the rubble. But whereas Martin responds primarily with an ethics of personal remembrance and collective survival – one that accepts that the initial event needs to be relived for the sake of all its victims – Saint-Éloi points towards an ethics of collective memory and rewriting of history that counteract earlier silencings of Haiti's recent and more distant past.

Time, temporality and narrative are critical concepts that have long been subject to literary-theoretical debate. At its most basic level, narrative can be described in terms of a story which has the core properties of an event or events, which proceeds chronologically through time, and which is conveyed through some representational medium.[4] A similar if somewhat more conceptual definition is given by Susan Onega and José Ángel García Landa, for whom narrative is 'the semiotic representation of a series of events meaningfully connected in a temporal and causal way'.[5] Story, temporality and causality are the key elements of narrative. Story is not the same thing as narrative, 'for where story events always proceed in chronological sequence, they can be narrated out of chronological sequence'.[6] This key differentiation, which is based on temporal difference between the occurrence of events and the ways in which they are accounted for, is shared by otherwise very different theoretical approaches to time and narrative.

Despite the lack of critical consensus on the myriad ways in which novels organise time, most theoretical frameworks of analysis tend to follow Gérard Genette's distinction between: *histoire* (story), for the mere chronology of events; *récit* (narrative), for the temporal order in which they are arranged; and *discours* (narrating), for the act of telling as represented or implied in the fiction.[7] In this framework, time can be compared to '*histoire*', 'story'; while temporality, the sum of narrative techniques for representing time, is equated with '*discours*', 'narrating'. Genette compares the relationship between the two under the categories of order, duration and frequency.[8] Order refers to the relation between 'the chronological sequence of the story events and the sequence in which they appear in the discourse'; duration denotes 'the relation between the length of time an event takes and the amount of space given to it in the novel' and frequency is the 'relation between the number of times an event occurs and the number of times it is narrated'.[9] Narrative time can also be characterised in terms of anti-transience, anti-sequence and anti-irreversibility. Narratives such as the two *récits* analysed here, when understood in relation to the transience of successive moments measured against objective time, i.e. a regular sequence of past-present-future, offer numerous techniques to recall the past, to make it present again, and to elongate, suspend and disrupt the presentation and experience of the passing of time. In the aftermath of the 2010 disaster, linear perceptions of time, i.e. those divided into clearly distinguishable concepts of past, present and future, are directly challenged by the traumatic experience of the earthquake, which inhabits all of these temporal categories at once while also inaugurating a new division of time into 'before' and 'after'.

Moreover, to a varying extent, theirs is a work of narrative reconstruction marked by both temporal and spatial displacement: Martin composes his account in the week following the earthquake as he moves between Fort-de-France (Martinique) and Paris (France), whereas Saint-Éloi completes his in June 2010 in Sauve (Canada). This remove from the 'here and now' of the January disaster adds a layer of mediation to the texts that raises wider questions about the privileged position of two non-resident writers whose formal status, in this particular context, would help to save their lives. At the same time, their texts also complicate easy binaries between 'insider' and 'outsider' perspectives – a categorisation which often equates the physical proximity or national belonging of the author with the authenticity, accuracy and ethical value of the text. Such simplistic divisions are inadequate reflections of the experience of displacement that is shared by many writers and intellectuals from Haiti and the wider Caribbean. Nor do they capture the complexity of affective and material ties, 'the impulse toward community with the simultaneous tensions of differences', between Haiti and the *Dizyèm Depatamen* (10th Department)[10] – a neologism for the Haitian *dyaspora* (diaspora) – and 'the world of differences that constitute the experience of daily life'[11] in the *dyaspora*. The Kreyòl term, as used by Claudine Michel, Brinda Mehta, the journalist Jean Dominique and others, has clear political connotations. It expresses 'the distinctiveness of ['the Haitian writers'] experience in the *dyaspora* as a conduit to negotiate identity and the parameters of belonging' and captures '[the] unique experience conditioned by the trajectory of experience'.[12] In effect, these networked relationships and processes of negotiation they demand cannot be confined to the dynamic of 'movement from one country or nation-state to another'[13] and involve 'the extending, developing, negotiating, and redefining of relations'.[14] Similarly, one's 'foreign' or 'insider' position, as occupied by some of the writers I bring together in this and subsequent chapters, is mediated and constructed across many levels of belonging and is shaped by histories of affective ties, sustained relations and the sense of obligations to and responsibility for a place, a community and those for whom one cares.

Saint-Éloi, who left Haiti for Canada in 2001, writes mainly from the position of a diasporic writer as 'semi-insider'.[15] Martin's life is also characterised by displacement, if of a different kind: originally from France, he lived for many years in Germany and Morocco before moving to Martinique – still a French overseas department and thus the European Union's outermost region – where he currently resides. Both writers travelled to Haiti to participate in the literary festival *Les Étonnants Voyageurs*, which was due to take place in Port-au-Prince from 14–17 January 2010. Martin was in Haiti on one of his regular visits to the island, while for Saint-Éloi this was an occasion to meet some of his fellow writers (for example Dany Laferrière, whose work is discussed in the next section) and to visit family relatives who still lived there. The earthquake struck while Saint-Éloi and Laferrière were having lunch at Hotel Karibe, in uptown

Port-au-Prince, soon after the two authors' arrival in the city. Martin's *Le Trem-blement: Haïti, le 12 janvier 2010* (henceforth *Le Tremblement*) and Saint-Éloi's *Haïti, kenbe la! 35 secondes et mon pays à reconstruire* (henceforth, *Haïti, kenbe la!*) are wide-ranging accounts of this violently interrupted journey. Neither was badly hurt, at least physically, in the earthquake. This sense of the arbitrariness of survival, partly due to their location, and the resulting obligation to witness it on behalf of those who died, are central to their respective narrat-ives of this seemingly unending disaster.

Through their emphasis on discontinuity and non-linear time, *Le Tremble-ment* and *Haïti, kenbe la!* both operate as memoirs of the earthquake and its aftermath. The split category of memoir, which 'refers to writing as a process of note-taking and to a piece of writing as a finished product at the same time',[16] captures well the fractured nature of these two post-earthquake narratives. In contrast to autobiography's discursive unity, memoir is a hybrid form, represent-ing an uneasy ensemble of several conflicting discourses. This marked formal difference reflects the many ways of envisaging the self in these two genres; for against the subjectivity of autobiography, 'which is presumed to be unitary and continuous over time, memoirs (particularly in their collective form) construct a subjectivity that is multiple and discontinuous'.[17] Heterogeneity and discontinu-ity are likewise amplified in the respective narratives' treatment of time and tem-porality. Varying in the temporal scope of their engagement, and inhabiting a different relationship to Haiti and its history, the two accounts confront the diffi-culty of distinguishing between the earthquake and its aftermath while testifying to the impossibility of confining the traumatic event within standard narrative categories or limiting it to the scale and force of physical tremors. The poly-phonic and unfinished character of memoir is highly relevant to these two post-disaster narratives that freely merge stories, reminiscences and their own personal experiences along with those of other survivors, with the aim of creat-ing a mosaic-like narrative that is appropriate to the broken nature of events. The kind of narrative linearity that usually accompanies autobiographical accounts is an unattainable goal; the traumatising roar of *goudougoudou* (the unearthly sound made by the tremors) permanently disrupts any attempt to account for it as just one in a sequence of many life-events.

The earthquake's echoing presence gives an untimely character to both Mar-tin's and Saint-Éloi's narrative reconstructions of the disaster. The concept of 'untimeliness', usually associated with Friedrich Nietzsche (e.g. his *Untimely Meditations* (1873–6)), is brought into a contemporary context by Ross Cham-bers in his 2004 study *Untimely Interventions: AIDS Writing, Testimonial, and the Rhetoric of Haunting*. In Nietzsche's case, untimeliness denotes the mode of anachronism that he sees as characterising critical thinking. For the German philosopher, 'a genuinely critical form of thinking must be *unzeitgemäss* – untimely, inopportune, unmodern – because it demands a rejection of everything governed by fashion'.[18] As Daniel Breazeale suggests, whereas the conformist 'public opinion strives always to be "timely", a declared critic of the same will instead flaunt his deliberate "untimeliness"'.[19] Hence untimeliness is 'doubly anachronistic' as, in the first instance, it 'entails a perspective informed by the past' and, in the second, it is 'resolutely future-orientated, as can be seen from Nietzsche's overriding concern with cultural renewal and rebirth'.[20] Untimeliness as an example of temporal disorder – as anachronism – brings together the categories of past and future. Thanks to its dual temporal character, it avoids the rigidity associated with linearity and hierarchy that are central to many literary narratives.

Contemporary uses of the 'untimely', including Chambers' own, draw on an overriding sense of discontinuity and disruption of the linear relationship between conventional categories of time in the context of what he calls 'aftermath writing'. Aftermath writing signifies 'the breakdown of reassuring categories that place trauma and survival of trauma in separate compartments',[21] and testifies to the return and repetition of the initial trauma at a time when it is supposedly over. Untimeliness, in this context, denotes the impossibility of differentiating between the trauma of the event and its aftermath while expressing the difficulty of 'returning "from" trauma to untraumatized life'.[22] Untimeliness, which is at once a state of separation and an awareness of co-presence, defines then the dual state of survivorhood that the narrators of both *Le Tremblement* and *Haïti, kenbe la!* inhabit. Theirs is the experience of 'surviving trauma',[23] understood both as an affirmation of one's survival of a traumatic event and as a registering of the 'fact of the pain's surviving into the present, the fact that one has not survived it so much as one is (still) surviving it'.[24] As a consequence, aftermath narratives are accounts of 'surviving trauma' *and* of the experience of 'trauma's not being over when one wishes it to be in the past'.[25] Aftermath writing, Chambers suggests, acts as the 'after-math of an initial "math" (the etymological metaphor is of a second mowing of grass in the same season as the first): it is 'a repetition – in transformed guise – of the initial traumatic event'.[26]

Aftermath writing thus has a double character. On the one hand, it involves the exploration of survival as an experience of untimeliness, i.e. 'of a baffling experience of time as, conjointly, the separation of past and present and their continuing copresence'.[27] On the other, it constitutes 'an art of untimely intervention, seeking to introduce an awareness of untimeliness into a culture that prefers to live in time as if the past had no place in the present and did not haunt (i.e. inhabit) it'.[28] In effect, aftermath writing, as a process rather than a finished object or a defined destination, does not make claims to total control over haunting memories. Instead, it bears witness to '[the] aporia of art and to its pain. It does not say the unsayable, but says that it cannot say it'.[29] It is an attempt to respond to the demand of witnessing on behalf of the victims of the event, 'but in a mode other than talking (representing)'.[30] *Le Tremblement* and *Haïti, kenbe la!*, on both formal and thematic levels, are defined by this two-fold experience of untimeliness as a joint apprehension of temporal separation and co-presence synonymous with a state of *hauntedness*: 'experience of trauma as that which fails to end, but continues to repeat and to return, even when it is supposedly "over"'.[31] The twin processes of structuring and narrativising the disaster,

initially undertaken by the respective narrators in the hope of achieving some
sense of closure, become instead a re-enactment: a painful experience of reliving
the initial event.

Aftermath chronologies

The tensions between conventional chronology, the beginning and the end of the
earthquake as a geological phenomenon, and the subjective experience of time in
the aftermath of the disaster are first visible in the formal aspects of the two narratives. Titles, chapter breaks, temporal markers and epigraphs are all employed
in both accounts to signal attempts to control the recent traumatic experience.
Lionel-Édouard Martin's seemingly ordered diary-form in *Le Tremblement:
Haïti, le 12 janvier 2010*, for example, tries chronologically to recreate the narrator's first-hand experience of the earthquake and the three days following it.
However, the title of the work already contrasts the precise time and place of the
event with the undefined concept of 'le tremblement' – 'the tremors'. The multiple meanings of the word 'tremblement' undermine the initial impression of
precision and challenge any sense of a clearly identifiable narrative with a
marked beginning and end. 'Tremblement' in its dictionary sense combines continuity and interruption, and can simultaneously mean 'a rapid succession of
small concussions, small fluctuations'; 'swaying, rapid change' and 'sudden
variations in the height and intensity, under the effect of a sensation or
emotion'.[32] In fact, the earthquake ('le tremblement de terre') is only one of
many different kinds of both literal and metaphorical disruptions and tremors –
'tremblements' – in the text. The title thus brings to the fore the tension between
the idea of a singular temporal moment and a repeated or extended occurrence
that marks the subjective perception of the earthquake and its immediate aftermath throughout Martin's memoir.

The full title of Saint-Éloi's work – *Haïti, kenbe la! 35 secondes et mon pays
à reconstruire*[33] – also plays, albeit in a different way, with the notion of continuity and the contrasting brevity of the event which, according to the author,
lasted only 35 seconds. The title combines an exhortation to the country, 'Haiti,
never give up'!,[34] with an expression of Saint-Éloi's personal relationship, which
is all the more significant in the context of his own diasporic position, to Haiti.
In addition, the short duration of the disaster is contrastively juxtaposed with the
sense of continuity indicated by the verb 'reconstruire', which can mean 'to construct anew a building, a piece of art' or, alternatively, 'to subject a work to a
new plan, reassemble the components in a new way in order to recreate a
whole'.[35] Both meanings of the word accentuate the link between past and
present and suggest the process of reassembling existing elements. In this way,
the title of Saint-Éloi's account already implies that the event's unprecedented
impact, and the sense of new history it potentially inaugurates, cannot be separated from the country's longer past.

The chapter breaks in *Le Tremblement* and *Haïti, kenbe la!* also significantly
contribute to the creation of narrative temporality by indicating different ways of

framing the event and varying levels of retrospective re-ordering. Allowing the narrator 'to play games on a metafictional level',[36] chapter breaks permit him at the same time to 'expound at length on generic, aesthetic and metanarrative matters'.[37] In Martin's text, each entry is marked by a subsequent date ('Le 12 janvier', 'The 12th of January'; 'Le 13 janvier', 'The 13th of January' etc.), beginning with the fateful day the earthquake struck. This simple chronological ordering lends a sense of urgency to the account and gives the narrative a highly personal character while suggesting, at the same time, that this is a documentary narrative, a chronicle and a report from the scene of disaster. *Haïti kenbe la!*, unlike Martin's memoir, is divided into 12 chapters, each with its own descriptive title, for example 'The earth sounded *goudou-goudou* ... ',[38] which reveals the act of retrospective ordering in the light of new, previously inaccessible knowledge. Although the first chapter title establishes the earthquake as a definitive beginning, subsequent ones – for example 'Flavours of childhood', 'The days after'[39] – do not convey the linear succession of events that is suggested in *Le Tremblement*. Rather, they implicitly comment on the form of testimonial memoir as a mixed-mode collection of interweaving narratives: the narrator's account of the earthquake, his memories of childhood folk tales and personal reflections on the country's past. Varying in their construction of narrative temporality, the two accounts point to the limitations of both approaches: neither linear chronology nor retrospective ordering can fully capture the untimely experience of 'that thing', *bagay la*.[40]

Furthermore, through the form of his account as aftermath writing, Martin hopes to make visible the unfolding of events, from the initiating moment onwards, to those who have been denied the chance to witness them. In a metaphorical vein, he seeks to provide a causal link between the 'before' and 'after' of the earthquake, both for himself and the many friends and colleagues he lost in the disaster. For him, as well as for Chambers, aftermath writing has a clear personal and a collective dimension as well as strong religious connotations. In Chambers' words, aftermath writing registers:

> [an] urge to witness, to awaken those who sleep, and to reawaken them, with a message of extremity that has trouble getting through [...] but does not lose its power to interrupt, disturb, trouble, and remind the sleepers (an anamnesis or counter-forgetting in the strictest sense) of what they (we) had never ceased to know.[41]

This process of awakening is an act of breaking the 'haze' created by a sense of relative comfort – of 'being comfortably at home, well fed'[42] – and of being at a remove from the extremity of the disaster to which witness narratives testify. As such, aftermath writing is a manifestation of an individual urge to share an experience which, despite countervailing attempts, cannot be classified as 'over'. Chambers characterises this process of counter-forgetting as 'anamnesis', pointing at the same time to the potentially religious character of the act of witnessing. 'Anamnesis' comes from the Greek and has two seemingly contradictory meanings. In its wider, etymological sense, it refers to '[a] calling to mind, remembrance, or memorial'.[43] Yet in its theological context, it 'is not the mere mental recall of something past, over and done with, nor is it the fond recollection of something or someone absent'.[44] Rather, as used in the Christian liturgical context, it is the *making present* – an acknowledgement of Christ's real presence in the Eucharist.[45] This theological dimension to the act of recollection becomes apparent in Martin's work, which draws on requiem and frequently uses liturgical references.[46] In so doing, Martin hopes to transform his account into a prayer – both an offering for and a conversation with the victims of the disaster.

This narrative prayer the narrator wishes to compose is to be brief and consequently easy to memorise, being described as 'a series of words which, like a prayer, can be easily memorised'.[47] In this way, the narrativisation of events becomes an attempt to establish a dialogue between survivors and non-survivors of the disaster. In a footnote to the entry 'Le 15 janvier', the narrator summarises this conversation as follows:

> To write is always to converse with the dead and no writer can cut himself off from these dialogues [...]. I write with those to whom my work is dedicated, there, my *Tremors* [reference to the title of the work] – a phone call from beyond the grave.[48]

Writing is compared here to a conversation 'from beyond the grave',[49] thereby creating a link between both the living and the dead and forging a connection between the past, which is synonymous with destruction and loss, and the present, which is experienced as a state of survival. The text thus testifies to a relationship of 'double continuity, with the dead and the living, combined with a double separation, from the living but also from the dead, with whom one nevertheless identifies and who are, so to speak, one's closest kin, as is death itself'.[50] This constitutive tension between the tacit desire to separate oneself from the event and the acknowledged impossibility of doing so is ingrained in the fabric of Martin's narrative as he tries to contain the experience within a chronological narrative design.

In effect, the narrator's prayer-like invocation works against the aforementioned linearity that otherwise forms the main organising principle of the narrative and is seemingly confirmed by the finality of its closing credits: 'Fort-de-France-Paris the 16th of January–22nd of February 2010'.[51] Far from being a 'closed' occurrence in the past, the earthquake, once converted into a prayer, is relived repeatedly each time the supplication is uttered. The narrator is forced to re-immerse himself in the same traumatising events from which he seeks an insulating distance. This sense of continuity that the narrative has to confront and hopes to sustain is emphatically not a linear ordering with a clear beginning and end, as the chapter and closing dates might indicate. On the contrary, it is an untimely chronology in which it becomes impossible to differentiate between the initial event and its aftermath, and which consequently transforms the prayer in the text from a simple invocation into an elaborate requiem:

> I now had to search in my memory for the parts of the requiem and find phrases, loose at first, to then *animate* them into a sequence of movements, orchestrate them and transform the tremor into a sacred hymn [italics original].[52]

The transition towards this requiem form, in its original sense 'a solemn choral service for the dead sung in Roman Catholic Churches',[53] becomes a way of entering into a dialogue with the dead – one that breathes in, instils life and touches the soul (*anima* in Latin) – and is a means of addressing God and asking him: '[r]equiem aeternam dona eis, Domine' – 'Lord, grant them eternal rest.'[54] By offering a requiem for the dead that is his text, the narrator hopes that he can honour the disaster victims, giving them the funeral they never had while also claiming the closure and relief for which he himself yearns.

At the same time, his imperfectly crafted prayer is also a quiet and shy attempt to join the Haitian survivors in singing songs of sorrow and praise: 'A prayer, an act of thanksgiving? It is difficult to know. But the voices are there, present, unendingly dignified and powerfully beautiful.'[55] This chorus the narrator hopes to join fills the vacuum created by the irrevocable absence of loved ones. Yet far from providing closure, the requiem forces the narrator to revisit the experience of the disaster: *relieving* becomes *reliving*. But though they are painfully present in each verse and refrain, these rekindled memories paradoxically enable the survivors to form, as Martin experiences it, a non-hierarchical community, a polyphonic if at times discordant and disharmonius one, that asserts itself against the void. As long as the chorus lasts, the menacing tremors are disarmed and become reassuring vibrations echoing the prayerful 'vibrato full of humility and repentance'.[56]

Similarly, highlighting this dual character of aftermath writing, epigraphs in Martin's text contribute to the creation of narrative temporality in terms of disjointed chronology and moment of writing. They highlight the process of narrativisation as a link between the past, which is understood both as the experience of the disaster and in terms of wider national history, and the present of survivorhood that the narrators unwillingly inhabit. *Le Tremblement* opens with two epigraphs: one from Aimé Césaire's narrative poem *Cahier d'un retour au pays natal* (1939) (*Notebook of a Return to the Native Land*) and the other from part III, '*Conversation entre les dormeurs moribonds*' ('*Conversation between the dying sleepers*') of Jean-Pierre Duprey's *Trois feux et une tour* [*Three flames and a Tower*] (1950):

> *like the penetrance of an apocalyptic wasp.*[57]
>
> Aimé Césaire,
Cahier d'un retour au pays natal
[*Notebook of a Return to My Native Land*]

The first epigraph from Césaire powerfully evokes the theme of the apocalypse, yet places it within the context of resistance against defeatism and colonial domination. In calling forth this earlier literary work, *Le Tremblement*, similarly to Rodney Saint-Éloi and Kamau Brathwaite (see Part I), makes a connection between the struggle against historical and contemporary forms of oppression, suggesting that the strength to carry on can only come from the act of joining the community of survivors in their collective attempt to express an overriding sense of sorrow and loss. This, in turn, affirms the victims' presence and agency: 'The strength is not within us but above us in a voice piercing the night and the audience like the sting of an apocalyptic wasp.'[58] Against the background of deadening images of Haiti as a defeated nation, Martin's narrative points to the possibility of an empowered, yet anything but naïve, coming together of the affected collective, fashioned from the refusal to forget the event and its haunting impact. This affirmation of the survivors' agency and voice begins with a discursive shift and the forging of a narrative form that is rooted in this untimely experience of the earthquake.

In addition to evoking a specific literary tradition and legacy of Césaire's writings, the adjective 'apocalyptic' implies a set of literary and theological connotations that contribute to the untimely character of Martin's memoir. In everyday use, the word 'apocalypse' is most often employed as an adjective that 'refers to a set of ideas about the end of the world or the end of time in which the cataclysmic event breaks suddenly in upon an unsuspecting world, and results in massive destruction and disaster'.[59] Yet in its etymological sense as well as its biblical[60] and theological significance, the term has a much more complex significance, one that is not limited to images of annihilation and plight. It implies a meditation on the end of times as much as a vision of the second coming of Christ. Already the word's origin, the term comes from 'the Greek term *apokalupsis* [which] means – disclosure or unveiling',[61] suggests a transformative potential; for 'interest in the end is not part of the definition of an apocalypse, simply a common feature'.[62] While carrying obvious eschatological connotations, the term 'apocalyptic', as Stefan Skrimshire suggests, announces 'a revelation or disclosure of the spatial and temporal *transformation* of the world [and] not simply a prediction of the "end time"' (my emphasis).[63] In this dual context, the earthquake can be seen as initiating a transformative, yet anything but didactic (see Part IV), process which, far from denoting the 'end of times', leads to an entirely different world order. In this last sense – and as Martin uses it – the earthquake is a total event marking an irreversible change and an end to the world as he previously knew it.

The second epigraph, taken from Duprey's Surrealist play, is a dialogue between two characters about death, the absurd and the irreconcilable sense of void they are both experiencing:

TALAMÈDE: *And death?*

PHILIME: *Is a button I have in my navel and when I press it, the hour of death will announce itself.*

TALAMÈDE: *And the next world?*
PHILIME: *It is death seen side on, full face and sideways at the same time.*
TALAMÈDE: *And the joined hands?*
PHILIME: *Are a prayer left at a gallop, or rather a bird stitched in the sky.*[64]

Jean-Pierre Duprey, *Trois Feux et une tour*

The theatrical piece culminates with a scene of annihilation: a gunshot and an explosion of a firecracker can be heard, then the stage is plunged into darkness only to be revealed again, filled with rubble.[65] This sense of existential emptiness and obscurity, shared by Martin, is integral to Duprey's *œuvre*, the creative aspiration of which is summed up by André Breton in one phrase: 'Let there be darkness!'[66] Martin's post-earthquake narrative is at once an acknowledgement of the all-encompassing obscurity that lurks behind each word and an act of refusal to give in to that immobilising desperation. In their creative work, both Duprey and Martin still see a space where the fragmented self can be tentatively reassembled.[67] These themes of transformation, death and disaster in the opening pages of the book are central to the narrator's aim to create a hybrid narrative form, incorporating elements of dialogue and prayer, through which he can provide some limited insight into the event, illustrate its personal and collective impact and its ongoing aftermath, and commemorate those who died.

Similarly, Saint-Éloi's account contains metatextual epigraphs that signal the way in which the narrative intends to relate the earthquake to personal and collective history while acknowledging the inadequacy of past division of time into past, present and future. To a greater extent than *Le Tremblement*, *Haïti, kenbe la!* emphasises the empowering potential of the text, which has the capacity to become a historical artefact for future generations of Haitians. The first epigraph is by a Haitian American author, Edwidge Danticat: 'We are not used to letting grief reduce us to silence.'[68] The second is an African proverb: 'As long as the lion will not have its historian, the stories of hunting will always glorify the hunter.'[69] Pointing to the importance of diasporic testimony and the historical dimension of the account, the two epigraphs suggest that the unfolding process of narrativisation offers the chance both to *reclaim* and to *revise* the historical narratives of Haiti. In this context, *Haïti, kenbe la!* is an attempt to assert a strong individual and collective voice in the aftermath of the earthquake and to compose an empowering narrative of the country's recent past. Rather than simply underlining the commemorative nature of the account, the epigraphs indicate that this narrative, in some ways similarly to notions of postcolonialism as a process and an anticipatory discourse (Part I), is future-oriented and has a collective importance. The two texts thus both attempt, in their different ways, to create a bridge between the past of the event and the present state of survivorhood that the narrators inhabit, while also envisaging possible futures in the wake of the earthquake. Their form has an untimely cadence that strikes against the silencing discourses on Haiti's past, present and future.

Timing disaster

But in the aftermath of the disaster, it is no longer possible to separate these three temporal categories or to present the unfolding of events in terms of a chronological sequence that relies on sustaining clear divisions between the present and the past. Instead, the overwhelming experience has its own untimely chronology. The various formal ways in which the texts engage with categories of time and narrative temporality foreshadow the key tensions that permeate them át the thematic level, while pointing directly to the dissonance between the personal experience of that extended and ongoing disaster, long after the ground stopped shaking, and the tremors' perceptible duration. At one level, *Le Tremblement* and *Haïti, kenbe la!* seek to present the disaster as a seemingly punctual event that can be measured using the standard units of objective clock-time. The short duration of the event, indicated already in the title as 35 seconds, is recalled throughout Saint-Éloi's account. Reflecting on its brevity, the narrator remarks:

> Thirty-five seconds.
> Thirty-five seconds.
> And everything shakes with the earth.
> Thirty-five seconds of devastation.[70]

The visual presentation of the lines, which appear after consecutive breaks of two lines in the text, reflects the symbolic significance of time measurement and reinforces the portrayal of the earthquake as a moment of rupture.

Yet the emphasis on the measurable length of the event, the 35 seconds of its duration, suggests that its end can be easily identified and confined within a defined time span. Revisiting the events of 12 January, the narrator insists that they have ended in a clear-cut manner. As he notes:

> The earth is restless and everything around spins.
> Thirty-five seconds.
> I look to my left to get a sense of direction in case the wall of the hotel fell. Suddenly *goudou-goudou* stops. In a sharp and clear manner.[71]

The sense of dizzying confusion that is expressed at the beginning of the passage contrasts with the temporal precision suggested in the following sentence, where the earthquake is again presented as a precise and finished occurrence. What *is* definite is the impact the brief tremors have had, instantly transforming the relationship between self and other, the individual and the national collective and their joint history. Employing diction and imagery rich in historical references, the narrator remarks:

> In thirty-five seconds everything changes for each of us and for the whole country. I am an other. Is identity always declined in the plural? I am not the

same person who, newly-landed, wanted to discover his native island. I looked alarmed. Desperate. [...] In a general disaster there is always a disaster of oneself.[72]

Here, the narrator differentiates between, but does not categorically separate, the exploration of this redefined subjectivity from the historical context of his enquiry. The language and imagery of the remark point to the personal experience of emigration while also hinting at the history of colonisation that underlies it. The earthquake, as an environmental hazard and as an untimely experience, is interpreted both as a 'general disaster' and 'a disaster of self'.[73] In accordance with these dual dimensions, it remodels the sense of personal as well as collective past and future. While admitting that everything has changed for each individual as well as for the country as a whole, the narrator still struggles to see himself as part of the national collective united by the shared experience of the disaster. Rather, he now perceives himself as 'a being at the bottom of the abyss, "an other" ',[74] and his earlier expressions of enthusiasm and communal fraternity are replaced by a sense of overwhelming despair and fright. In a similar fashion, the precision he initially attributes to the earthquake is later revised, and he is no longer certain how long the initial tremors lasted: 'The earth trembled in less than a minute. The people trembled with it. And everything turned into dust.'[75]

The changing and life-defining significance of a minute is similarly recalled throughout *Le Tremblement*. A minute, as a standard measure of time, becomes synonymous with the event, and the ordinary unimportance of one minute in the narrator's previous life is contrasted with the extraordinary significance of each second in the context of the earthquake. Acknowledging this drastic shift, the narrator contemplates:

A minute does no more than round things up. It's not a big thing in the life of a man. But, sometimes, there are minutes which expand their seconds to the extreme conferring on them the layers of rubble of death.[76]

Minutes and seconds, 'expanded to the extreme',[77] are no longer neutral measures of temporality, but rather become irresistible reminders of the disaster; from now on they can determine life or death. The shattering experience of the tremors, like the material devastation of the city, is simply too overwhelming to comprehend. As the two narratives ultimately reveal, the temporal limits that are ostensibly imposed on the earthquake can in no way measure the real impact of the event.

Similarly, everyday objects, such as photographs and personal travel accessories, function, in varying ways, as signs of Martin's own survival but also as mnemonic triggers, becoming untimely souvenirs of the experience – tokens of remembrance[78] – that displace him from the 'here and now' to the experiential site and time of the disaster. In Martin's memoir, the photographs of destruction with which the narrator is presented cause an experience of temporal disjunction that forces him to acknowledge his inability to control the circumstances surrounding the event. In looking at a series of images of urban devastation that have been given to him by a journalist, the narrator in *Le Tremblement* is re-immersed in the original traumatic experience, reliving it as a series of haunting repetitions. Commenting on the disturbance these images create, he writes:

> It's a long computing sequence of destructions. The viewing of each photo does not last more than few seconds allotted by the electronic circuit – always the same few seconds regardless of the photo. But the duration of these images is not the same as the duration of *my* images [italics original].[79]

The passage reveals a dichotomy between prior *knowledge* of the duration of time, which is measured in seconds and minutes, and the contrasting *experience* of temporality as a conflation of the narrator's past memories, accumulated over ten years of regular visits to Haiti, with present scenes of post-earthquake devastation.

Moreover, the sequence in which the photographs are displayed to the narrator has a cumulative rather than ordering effect in so far as the images keep referring to the original moment the earthquake struck. In Susan Sontag's terms, the photographs present 'a neat slice of time, not a flow',[80] which reaffirms the status of the event as a redefining moment in the narrator's life. Consequently, the years preceding the earthquake seem to speed up, compressing the gap between 'before' and 'after' the disaster:

> The ten years during which I visited Haiti regularly become undistinguisha-ble, they follow one another more quickly than in reality and in a cross-fade manner, like time-lapse film of a blossoming flower.[81]

In this passage, the narrator evokes cinematic effects of time-manipulation and superimposition, i.e. when two or more images are placed over each other in the frame, to distil the general sense of temporal confusion into a single image. Again, the temporal control he hopes to sustain cannot withstand the over-whelming nature of the experience, and he finds himself no longer able to recall the individual sites as he had remembered them before the disaster struck. Rather, hidden under shattering images of debris, the memories of these sites function as ghostly reminders of an inaccessible past. The photographs redefine the way in which the narrator interprets his past and records his memories of events and places with each one triggering analepsis, a resurfacing of earlier associations and memories that merge inexorably with the disaster. The images reproduce and repeat, in mechanical form, that which 'could never be repeated existentially',[82] thereby forcing the narrator to re-encounter the destruction and personal loss caused by *goudougoudou*. As souvenirs of the traumatic rupture, the images of the ruined capital '*fill the sight by force*' [italics original] and vio-lently override previous memories of Port-au-Prince.[83]

Likewise, trivial personal objects also become powerful reminders of a time and place irreparably gone-by yet present as a haunting memory and traumatic wound. The shampoo and soap, that the narrator of *Le Tremblement* collected from Hotel Karibe, disrupt the linear perception of time, undermining his ability to mark an end to the disaster and thus achieve a sense of closure. Attempting to detach himself from the scene of the catastrophe, on two occasions, the narrator writes:

> That's what power is: to have those little things, master them, render them powerless – in the same way we disarm grenades so that they can't explode anymore – and through them establish a bridge between here and *there*.[84]
>
> The soap and the shampoo from Karibé from *before*. I understood the pleasure I had in using them the night before; they emanated *before* and created a faultless continuity of time and space [italics original].[85]

His hope is that physical ownership of the objects from the hotel will give him a sense of mastery over his memories and the associations they can potentially trigger. However, as the two passages reveal, such momentary assertions of control are continually thwarted. The military comparison the narrator makes between 'those little things' and 'grenades'[86] suggests that, despite his earlier claims, the items still provoke overpowering memories of the event. The original verb 'émaner' (emanate) used in the second passage, which means 'to emanate, to originate',[87] signals the object's rootedness in the past, conceived here as the time '*before*' the earthquake struck. The disaster again becomes a total reference point that redefines retrospectively even the most seemingly trivial reminders of the narrator's life before 12 January. Similarly, the expression 'sans faille' (faultless), which conveys a sense of undisturbed continuity, also draws attention to the rift the narrator is attempting to heal and the time 'without breaks' for which he is melancholically longing.

The italicised deixis, '*là-bas*' ('*there*')[88] also implicitly refers back to the site of the disaster – a place so traumatic he is unable to name it. Deixis provides a useful strategy, a 'linguistic means by which the speaker anchors utterances in the concrete place of enunciation ("here", "there", "this table")'.[89] The narrator uses temporal and spatial deictic markers – 'here and *there*', 'ici et *là-bas*',[90] and 'before', '*avant*'[91] – to register shifts in narrative setting, exposing in the process his inability to name the original date of the disaster and to fix its primary location. The first of these refers to Guadeloupe, where the narrator temporarily resides; the second, whose importance is underlined by italics, to the site of destruction. Port-au-Prince, now associated only with the earthquake, becomes a 'place' in Jeff Malpas's phenomenological understanding of place as a site for the formation of subjectivity, one 'not founded *on* subjectivity, but […] rather that on which subjectivity is founded'.[92] The structure of subjectivity is 'given in and through the structure of place'[93] and the destruction of it, as well as any subsequent reconstruction, requires a redefinition of one's perception of self. Together, both seemingly objective measures of time as well as referents of time and place, acting as rhetorical substitutes for the disaster, become metonymies of the traumatic experience and trigger the process of cognitive and semantic association. Without being directly named, the earthquake, experienced as an ongoing immersion in a state of untimely survivorhood, is consistently present in the narratives' treatment of space and time, their form as aftermath writing, and is the defining spatial and temporal referent, long after the city stopped shaking.

Notes

1   In the time period since the earthquake's origin at 2010–01–12 21:53 to 2010–02–23 17:00 UTC, the USGS NEIC has located 59 aftershocks of magnitude 4.5 or greater. Sixteen of these aftershocks have magnitudes of 5.0 or greater. The two largest aftershocks were magnitude 6.0 and 5.9. The M 6.0 aftershock occurred 7 minutes after the main shock on January 12 and the M 5.9 event occurred at 11:03 UTC on January 20.

US Geological Survey, *Magnitude 7.0 Haiti*

2 In their attempt to comprehend traumatic experience, numerous post-earthquake narratives also employ religious diction which has complex theological connotations. Several of the texts analysed in the thesis exemplify the multiple uses of theologically resonant language. In *Le Tremblement*, for example, the vocabulary of regeneration and redemption, which is rooted in the concept of Messianic history, is employed to honour the deaths of the disaster victims. The narrator openly states that he wants to transform his account into a prayer (a claim analysed in depth later in this chapter), and throughout the text he engages with the religious character of the form. Another use of religious diction is exemplified in Dan Woolley's *Unshaken* – a survivor account which will be analysed in Part IV. Here, the narrator uses religious discourse to frame the experience of the disaster within a conversion narrative. The concepts of memory, remembrance, testimony and witness, which are key terms in relation to post-seismic texts, also have complex religious provenance. Kerwin Lee Klein emphasises that not enough attention is given to the theological origins of these terms, which are so prominent in current scholarship and popular discourse:

> In academic and popular discourse alike, *memory*, and its associated key words continue to invoke a range of theological concepts as well as vague connotations of spirituality and authenticity. Authors writing in secular academic contexts necessarily trade upon these associations but seldom make them explicit. Part of that trade stands upon the place of remembrance in Judeo-Christian tradition – '*Zakhor*' (remember) in the Old Testament, and 'Do this in remembrance of me,' in the New.
>
> Kerwin Lee Klein, 'On the Emergence of Memory in Historical Discourse',
> *Representations: Special Issue: Grounds for Remembering*, 69 (2000)',
> 127–50 (p. 130)

3 Here, I am referring to the title of Maurice Blanchot's *The Writing of the Disaster* (1980) and the ways in which the text can be productively re-read in the context of contemporary ecological crises, such as those directly preceding and shaping the January earthquake, since, in Joshua Schuster's gloss, 'to engage with contemporary ecology one must also engage with disaster' (Schuster, p. 164). The writing of the disaster contemplates the punctual yet embedded nature of disasters: '[d]isaster is not just loss, then, but an event that marks a catastrophic or irrevocable change in a system that was implicit in the system to begin with and that effectuates a new kind of system' (Schuster, p. 166). Narrative responses to the January earthquake, such as those analysed in Part III, attempt to grapple precisely with this composite character of disasters as an irrevocable change yet one that 'effectuates a new kind of system':

here, a complex post-earthquake island ecology. Joshua Schuster, 'How to Write the Disaster', *The Minnesota Review*, 83 (2014), 163–71 (pp. 164, 166).

4 H. Porter Abbott, 'Narrative', in *The Encyclopedia of The Novel*, ed. by Peter Melville Logan (Oxford: Wiley Blackwell, 2011), pp. 533–40 (p. 533).

5 Susan Onega and José Ángel García Landa, *Narratology: An Introduction* (London: Longman, 1996), p. 3.

6 Abbott, p. 534.

7 Catherine Gallagher, 'Time', in *The Encyclopedia of The Novel*, pp. 811–17 (p. 813).

8 James Phelan, 'Narrative Technique', in *The Encyclopedia of The Novel*, pp. 549–53 (p. 550).

9 Phelan, p. 550.

10 Up until 2003 Haiti was divided into nine departments: Sud-Est, Sud, Ouest, Nord-Ouest, Nord-Est, Nord, Grand' Anse, Centre Hinche and Artibonite. The Diaspora was then referred to as 'the Tenth Department'. In 2003, an additional department was designated in mainland Haiti, Nippes. Following violent manifestations and road blockages in September 2014, there have been calls for a further decentralisation and an increase in the number of departments from 10 to 14 or even 16. As of 2015 there are still ten departments. Frantz Duval, 'Editorial', *Le Nouvelliste*, 29 September 2014 http://lenouvelliste.com/lenouvelliste/article/136410/Le-nouvel-ordre-geographique-et-administratif-dHaiti.html [accessed 17 November 2014].

11 Nina Glick Schiller, 'Forward: Locality, Globality and the Popularization of a Diasporic Consciousness: Learning from the Haitian Case', in *Geographies of the Haitian Diaspora*, ed. by Regine O. Jackson (New York: Routledge, 2011), pp. xxi–xxix (p. xxii).

12 Brinda Mehta, *Notions of Identity, Diaspora, and Gender in Caribbean Women's Writing* (New York: Palgrave Macmillan, 2009), p. 14, in Claudine Michel, *Kalfou Danje*, p. 194.

13 Karen Fog Olwig, *Caribbean Journeys: an Ethnography of Migration and Home in Three Family Networks* (Durham and London: Duke University Press, 2007), p. 10.

14 Olwig, p. 9.

15 Graham Huggan usefully employs Rob Nixon's concept of the 'semi-insider' to analyse travel narratives and the ambivalent position of the narrator within these works:

> In classically Orientalist travel narratives, the narrator operates as an enabling interlocutor, usually for the purpose of peddling stereotypes under the guise of supplying cultural information to the West. A variant on this is the type of account produced by what Rob Nixon calls a "semi-insider", such as someone born and bred in the Orient, but reporting back on it from and for the West.
> Rob Nixon, *London Calling: V. S. Naipaul, Postcolonial Mandarin* (New York, NY: Oxford University Press, 1992), p. 28, in Graham Huggan, *Extreme Pursuits: Travel/Writing in an Age of Globalization* (Ann Arbor, MI: University of Michigan Press, 2009), pp. 162–3

16 Julie Rak, 'Are Memoirs Autobiography? A Consideration of Genre and Public Identity', *Genre*, 36 (2004), 305–26 (p. 319).

17 Lee Quinby, 'The Subject of Memoirs: The Woman Warrior's Technology of Ideographic Selfhood', in *De/Colonizing the Subject: The Politics of Gender in Women's Autobiography*, ed. by Sidonie Smith and Julia Watson (Minneapolis, MN: University of Minnesota Press, 1992), pp. 297–321 (p. 298).

18 Helen Groth and Paul Sheehan, 'Introduction: Timeliness and Untimeliness', *Textual Practice*, 26 (2012), 571–85 (pp. 571–2).

19 Friedrich Nietzsche, *Untimely Meditations*, trans. by R. J. Hollingdale (Cambridge: Cambridge University Press, 1997), p. xlvi.

20 Groth and Sheehan, p. 572.

21 Ross Chambers, *Untimely Interventions*, p. xxii.
22 Chambers, p. xxii.
23 Chambers, p. xxi.
24 Chambers, p. xxi.
25 Chambers, p. xxi.
26 Chambers, p. xxi.
27 Chambers, p. 191.
28 Chambers, p. 191. Here and elsewhere 'culture' is used in both the broad sense of every day human activities, the daily practices to live one's life, as well as to denote 'certain crucial instances of aesthetic reception', including modes of artistic representation. Chambers, p. xiii.
29 Jean-François Lyotard, *Heidegger and 'the Jews'*, trans. by Andreas Michel and Mark S. Roberts (Minneapolis, MN. and London: University of Minnesota Press, 1997), p. 47.
30 David Caroll, 'The Memory of Devastation and the Responsibilities of Thought: "And let' s not talk about that" ', in *Heidegger and 'the Jews'*, pp. vii–xxix (p. xi).
31 Chambers, p. 190.
32 '[s]uccession rapide de petites secousses, de petits mouvements d'oscillation'; '[v]acillement, variation rapide d'intensité'; and '[b]rusque variation dans la hauteur, l'intensité, sous l'effet d'une sensation, d'une emotion' in 'tremblement', in *Le Trésor de la Langue Française Informatisé* (2012) www.cnrtl.fr/definition/tremblement [accessed 08 December 2012]. [Unless otherwise stated all translations are my own.]
33 *Haiti, Haiti, Hold on Tight* [*or Don't Give Up*]! *35 Seconds and My Country to Rebuild*. [Unless otherwise stated all translations are my own].
34 Literally, 'Haiti, withhold'.
35 '[c]onstruire de nouveau un édifice, un ouvrage d'art' or, alternatively, '[s]oumettre une œuvre à un nouveau plan, assembler des éléments de manière différente pour refaire un ensemble' in 'reconstruire', in *Le Trésor de la Langue Française Informatisé* (2012) www.cnrtl.fr/definition/reconstruire [accessed 08 December 2012].
36 Monika Fludernik, *An Introduction to Narratology*, trans. by Patricia Häusler-Greenfield and Monika Fludernik (London and New York, NY: Routledge, 2009), p. 24.
37 Fludernik, p. 24.
38 'The earth sounded *goudou-goudou* …' Le terre a fait *goudou-goudou*', [The title of Chapter 1].
39 'Flavours of childhood', 'The days after'; Les saveurs de l'enfance', 'Les jours d'après'.
40 See Part I for the discussion of the use of the term 'that thing' (*bagay la* in Haitian Kreyòl) for the 2010 January earthquake.
41 Chambers, pp. viii–ix.
42 Chambers, p. viii.
43 Allan Bouley, 'Anamnesis', in *The New Dictionary of Theology*, ed.by Joseph A. Komonchak, Mary Collins, and Dermot A. Lane (Dublin: Gill and Macmillan, 1987), pp. 16–17 (p. 16).
44 Bouley, in *The New Dictionary of Theology*, pp. 16–7 (p. 17).
45 Paul VI, 'Constitution on the Sacred Liturgy Sacrosanctum Concilium', (1963) www.vatican.va/archive/hist_councils/ii_vatican_council/documents/vat-ii_const_19631204_sacrosanctum-concilium_en.html (para. 102 of 130) [accessed 08 Deceember 2012].
46 These include: the use of diction with strong religious and liturgical resonance such as 'le saveur', notre saveur' ('the saviour', 'our saviour' (p. 71)); references to Christ's Last Supper, 'la cène' (p. 107) and the reference to Jesus's act of breaking the bread and sharing the cup of wine with his disciples: 'Prenez et mangez-en tous […]. Car ceci est mon corps […]. Car ceci est mon sang' ('Take, eat; this is my body' and 'for this is my blood' (Mt, 26:26–29) (pp. 65–6). Both the breaking of bread and the Last Supper are signs that anticipate the sacrament of the Eucharist,

'an action of thanksgiving to God [which] recalls the Jewish blessings that proclaim […]. God's works: creation, redemption and sanctification' (*Catechism of the Catholic Church*, p. 298, para. 1328). In this particular context, these references add to the character of the memoir as a prayer and an offering for the victims of the disaster that, in contrast to Woolley's narrative (Chapter Three), refuses to be centred on the narrator's own suffering. *Catechism of the Catholic Church* (London: Chapman, 1994), p. 298.

The latter references, which follow the biblical wording and that of the Catholic Mass, are used in a dialogue between the narrator, Dany Laferrière, and Rodney Saint-Éloi as they share the food they managed to buy from a shop that was still open.

47 'une succession résumée, allant à l'essentiel, comme est résumé, resserrée, sur quelques paroles, la prière jamais si longue qu'on ne puisse la connaître par cœur'. Lionel-Édouard Martin, *Le Tremblement*, p. 121.

48 For him:

> To write is always to converse with the dead and no writer can cut himself off from these dialogues […]. I write with those to whom my work is dedicated, there, my *Tremor* [reference to the title of the work] – a phone call from beyond the grave.
>
> Martin, p. 118

'Écrire, c'est toujours converser avec des morts, aucun écrivain ne peut s'abstraire de ces dialogues […]. Moi, j'écris avec mes dédicataires, là, mon *Tremblement* – tout droit tiré d'un coup de téléphone et d'outre-tombe.'

49 Martin, p. 118.

50 Chambers, p. xxiv.

51 Fort-de-France-Paris 16 janvier–22 février 2010, Martin, p. 131.

52 'Il me fallait, maintenant, chercher dans ma mémoire les éléments du requiem, trouver les phrases, d'abord disjointes: puis les *animer* d'enchaînements, puis les orchestrer – faire, du tremblement, un chant sacré', Martin, p. 121.

53 'requiem', in *World Encyclopaedia Online* (Philip's, 2012) http://0-www.oxford reference.com.wam.leeds.ac.uk/view/10.1093/acref/9780199546091.001.0001/acref-9780199546091-e-9811?rskey=NFhIGO&result=1&q=requiem [accessed 01 February 2013].

54 John Bowker, 'requiem', in *The Concise Oxford Dictionary of World Religions Online* (Oxford University Press, 2012) www.oxfordreference.com.wam.leeds.ac.uk/view/10.1093/acref/9780192800947.001.0001/acref-9780192800947-e-6057 [accessed 01 February 2013].

55 'Prière, action de grâce? Il est difficile de savoir. Mais les voix sont là, présentes, interminablement dignes, puissamment belles.' Martin, p. 38.

56 'Cela dure, et la terre est dure, que nous endossons à plein corps, et qui, désormais, ne tremble plus mais vibre, à l'unisson des voix, d'un vibrato comme humble et chargé de repentir.' Martin, p. 38.

57 '… *comme la pénétrance d'une guêpe apocalyptique*'. The words come from the following passage in Césaire's *Cahier d'un retour au pays natal*:

> Et nous sommes debout maintenant, mon pays et moi, les cheveux dans le vent, ma main petite maintenant dans son poing énorme et la force n'est pas en nous, mais au-dessus de nous, dans une voix qui vrille la nuit et l'audience comme la pénétrance d'une gûepe apocalyptique. Et la voix prononce que l'Europe nous a pendant des siècles gavés de mensonges et gonflés de pestilences, car il n'est point vrai que l'œuvre de l'homme est finie que nous n'avons rien à faire au monde/que nous parasitons le monde/qu'il suffit que nous mettions au pas du monde/mais l'œuvre de l'homme vient seulement de commencer/et il reste à l'homme de conquérir toute indirection immobilisée/aux coins de sa ferveur/et aucune race ne

possède le monopole de la beauté, de l'intelligence, de la force/et il est place pour tous au rendez-vous de la conquête et nous savons maintenant que le soleil tourne autour de notre terre éclairant la parcelle qu'à fixée notre volonté seule et que toute étoile chute de ciel en terre à notre commandement sans limite.

And now we are standing, my country and I, hair in the wind, my hand small now in its enormous fist and strength is not within us, but above us in a voice piercing the night and the audience like the penetrance of an apocalyptic wasp. And the voice pronounces that Europe has been stuffing us with lies and bloating us with pestilence for centuries,/ for it is not true that man's work is completed/ that we have nothing to do in the world/ that we parasite the world/ that all we need is to walk in step with the world/ but man's work has only begun/ and man has yet to conquer every prohibition paralysed in the corners of his fervour and no race holds a monopoly of beauty, of intelligence, of strength/ and there is room for all at the rendez-vous of conquest and now we know that the sun revolves around our land shining over the plot chosen by our will alone and that every star falls from sky to earth at our limitless command.

> Aimé Césaire, *Notebook of a Return to My Native Land/ Cahier d'un retour au pays natal*, trans. by Mireille Rosello with Annie Pritchard (Newcastle upon Tyne: Bloodaxe Books, 1995), pp. 124–7

58 Aimé Césaire, *Return to My Native Land*, trans. by John Berger and Anna Bostock (Baltimore: Penguin Books, 1969), p. 30. Some translations prefer to use 'penetrance' for the French 'pénétrance'. See, for example, Aimé Césaire, *Notebook of a Return to the Native Land*, trans. by Clayton Eshleman and Annette Smith (Middletown, CT: Wesleyan University Press, 2001), p. 44.

59 Christopher Rowland and John Barton, eds., *Apocalyptic in History and Tradition* (Sheffield: Sheffield Academic Press, 2002), p. 1.

60 In the more specific context of biblical studies, it 'refers to a specific genre of text linked predominantly to Jewish and Christian traditions', exemplified in the New Testament Book of Revelation. Stefan Skrimshire, 'How Should We Think About the Future', in *Future Ethics: Climate Change and Apocalyptic Imagination*, ed. by Stefan Skrimshire (London: Continuum, 2010), pp. 1–11 (p. 4).

61 Rowland and Barton, p. 1.

62 Rowland and Barton, p. 1.

63 Skrimshire, p. 4.

64   T: And death?/ P: Is a button I have in my navel and when I press it, the time/hour of death will announce itself. /T: And the next world?/P: It is death seen side on, full face and sideways at the same time. T: And the joined hands?/P: Are a prayer left at a gallop, or rather a bird stitched in the sky

>   (Unless otherwise stated, translations are my own)

> TALAMÈDE: *Et la mort?*
> PHILIME: *Est un bouton que j'ai au nombril, et si j'appuie dessus, l'heure-cadavre va s'annoncer.*
> TALAMÈDE: *Et l'au-delà?*
> PHILIME: *Est la mort vue de haut, de face, et de profil, en voyant de côté.*
> TALAMÈDE: *Et les mains jointes?*
> PHILIME: *Sont une prière partie au galop, ou plutôt un oiseau cousu dans le ciel.*

65 '(*Après un coup de pistolet et l'éclair d'un petard, la scène s'obscurcit et s'emplit de décombres. Salex, Talamède et Quicri-Kirat deviennent les héritiers des propriétaires de la Mort.*)' Jean-Pierre Duprey, 'Trois feux et une tour', in *Derrière son double: Œuvres complètes*, ed. by François Di Dio (Paris: Gallimard, 1999), pp. 49–59 (p. 57).

66 'Que la ténèbre soit'. The phrase coined by Breton to characterise Duprey's work is an obvious inversion of the biblical invocation 'Let there be light!' from Genesis 1.3. André Breton, 'Préface', in *Derrière son double: Œuvres complètes*, ed. by François Di Dio (Paris: Gallimard, 1999), pp. 23–7 (p. 25).

67 This unifying potential of writing is also emphasised by Anne Bourgain in her commentary on Duprey's work. Following the traumatic experience of having to bury the dismembered victims of the bombardment of Rouen, Duprey's sense of self involves fragmentation and disjunction between mind and body. He is 'toujours plongé dans un monde étrange, fragmentaire'; his writing 'tente de recoudre les morceaux d'un moi disloqué'. Anne Bourgain, 'L'autre en soi ou la question du double chez un poète,' *Le Coq-héron*, 1 (2008), 97–104 (p. 97).

68 'Nous n'avons pas l'habitude de laisser notre chagrin nous réduire au silence.' (Edwidge Danticat). Saint-Éloi, opening epigraph.

69 'Tant que le lion n'aura pas son historien, les histoires de chasse glorifieront toujours le chasseur.' Saint-Éloi, opening epigraph.

70 'Trente-cinq secondes./Trente-cinq secondes./Et tout tremble avec la terre./Trente-cinq secondes de saccage.' [Double spacing in the original]. Saint-Éloi, p. 18.

71   La terre a la bougeotte, et tout autour pirouette.
    Trente-cinq secondes.
    Je regarde du côté gauche pour m'orienter au cas où les murs de l'hôtel tomberaient. Soudain, le *goudou-goudou*s'arrête. Net.

Saint-Éloi, pp. 34–5

72 'En trente-cinq secondes, tout change pour chacun d'entre nous, et pour le pays. Je suis alors un autre. L'identité se décline-t-elle toujours au pluriel? Je ne suis plus celui que, fraîchement débarqué, voulait redécouvrir l'île natale. J'ai un air plutôt effaré. Désespéré. [...] Dans un désastre général, il y a toujours le désastre de soi.' Saint-Éloi, p. 38.

73 Saint-Éloi, p. 38.

74 Saint-Éloi, p. 38.

75 'La terre a tremblé en moins d'une minute. Les hommes ont tremblé avec. Et tout est devenu poussière.' Saint-Éloi, p. 104.

76 'Qu'est-ce qu'une minute, celle qui fait le compte rond? Pas grand-chose dans la vie d'un homme. Mais il est, quelquefois, des minutes qui dilatent à l'extrême leurs secondes et leur confèrent une épaisseur de gravats et de mort.' Martin, p. 120.

77 '[des minutes]. qui dilatent à l'extrême leurs secondes'. Martin, p. 120.

78 'souvenir', in *Oxford English Dictionary Online* (Oxford University Press, 2013) www.oed.com/view/Entry/185321?rskey=0KQdXq&result=1&isAdvanced=false#eid [accessed 12 April 2013].

79 'C'est une lente suite informatique de destructions. Chaque photo ne dure, dans son visionnement, que les quelques secondes imparties par des circuits électroniques – toujours les mêmes quelques secondes, quelle que soit la photo. Mais la durée de ces images n'est pas celle de *mes* images.' Martin, p. 85.

80 Susan Sontag, *On Photography* (London: Penguin Books, 2002), p. 17.

81 'Ces dix années que je me suis rendu régulièrement en Haïti se superposent, se succèdent, en une façon de fond enchaîné où, bien plus vite que dans la réalité, l'avant se métamorphose en l'après, comme on filme en accéléré l'épanouissement d'une fleur.' Martin, pp. 85–6.

82 Roland Barthes, *Camera Lucida: Reflections on Photography*, trans. by Richard Howard (New York, NY: Hill and Wang, 1981), p. 4.

83 Barthes, p. 91.

84 'C'est ça, la force: posséder ces petites choses, les maîtriser, rendues inoffensives – comme on désamorce des grenades, et dès lors elles ne peuvent plus exploser – et par leur entremise établir une passerelle entre ici et *là-bas*.' Martin, p. 111.

85 'Le savon, le shampooing du Karibé *d'avant.* J'ai compris le plaisir que j'avais eu, la veille, à les utiliser: parce qu'ils émanaient *d'avant*, qu'ils créaient une continuité sans faille dans l'espace et le temps.' Martin, p. 113.

86 Martin, p. 111.

87 'émaner', in *Reverso: Collins Online English-French Dictionary* (HarperCollins Publishers, 2012) http://dictionary.reverso.net/french-english/émaner [accessed 22 November 2012].

88 Martin, p. 111.

89 Russell West-Pavlov, *Spaces of Fiction/Fictions of Space: Postcolonial Place and Literary Deixis* (Basingstoke: Palgrave Macmillan, 2010), p. 2.

90 Martin, p. 111.

91 West-Pavlov, p. 121.

92 In my analysis, I build on Malpas's understanding of the idea of place which encompasses both the idea of the social activities that are expressed through the structure of a particular place and the idea of the physical objects and events in the world that constrain, and are sometimes constrained by, those social activities and institutions. Malpas, *Place and Experience*, p. 35.

93 Malpas, p. 35.

5 Recalling history and reimagining the future

The disaster's untimely character, manifested in altered subjective perceptions of temporality, the passing of time and everyday objects that become referents of the event, also reveals itself in the ways in which history and the writing of 'new history' are envisaged in the two texts in the earthquake's aftermath. The personal sense of the earthquake as becoming an ever-present referent of time and space is also discernible in notions of collective experience of time and temporality following the 2010 January tremors. To a varying extent across the two works, the catastrophe haunts and halts the community of fortuitous survivors as they attempt to find anew their sense of the *here* and *now* in relation to a radically redefined understanding of individual and collective pre-earthquake past. As they contemplate these new histories, both *Le Tremblement* and *Haïti, kenbe la!*, albeit in different ways, suggest an ethics of writing rooted in the refusal, however painful, to mute the haunting echoes of *goudougoudou*.

In Martin's account, the event draws the individual into the collective, communicating a shared experience of survival in the context of proximity to death and devastating loss. As the narrator remarks:

> [...] there will never again be 'like before,' the breach, even sealed off and filled with plaster and mortar, will remain alive in the walls and in me; from now on I have definitively become a mature man with no hope of turning back.[1]

While annihilating previous temporal divisions, the earthquake thrusts the narrator into a painful maturity that is defined by an experience of frailty and powerlessness resulting from the sense of uncontrollable acceleration of time caused by the event and the sudden deaths of many of the narrator's friends and colleagues. Encapsulating the dual experience of break and of crushing overwhelm, the word 'mûr' (mature) becomes a homonym for 'mur' ('wall' or a 'barrier'). This unlikely pairing refers back to the devastation caused by the event as well as to the necessary detachment from previous personal history with the forced severance acquiring an almost visceral dimension that intensifies the pain of loss:

> Whatever we are leaving behind, it is a part of ourselves and our personal
> history we are getting rid of – in order to, if we survive this battle, be thrown
> into a new history.[2]

A sense of uncertainty and doubt permeates this passage with the plural pronoun
'nous', 'we', accentuating the collective nature of the experience. This community of experience is anything but a stable and homogenous one with its transience countered only by the permanence of the indelible imprint left in their
lives by the earthquake. The verb 'délestrer' ('getting rid of'), which means 'to
unballast', 'to relieve',[3] further adds to this ambiguity as it denotes both a separation and freeing from a burden – here, that of a traumatic past. Yet, however
hopeful he might be, the narrator is equally aware that any such attempt will
likely fail and that the 'new history' the event inaugurates will always be linked
to the disaster from which it springs.

Although *Haïti, kenbe la!* also points to the earthquake's redefining impact on
perceptions of time, self and other, it stresses more than does *Le Tremblement*
the significance of the disaster for collective perception of Haiti's national
history. The event is 'a fracture without precedent', ('[u]ne fracture sans
pareille'),[4] has turned the entire history of the country upside down, and a new
history has emerged, to be defined by and as from the moment the disaster
struck.[5] Directly responding to this experience of disrupted history, *Haïti, kenbe
la!* forges a composite narrative aesthetic that merges different accounts of the
earthquake in an attempt to create a multivocal narrative. This layered narrative
weaves between different 'memories of the times' which, in effect, challenge
previous notions of the past as both a temporal and a discursive category. These
recounted, overheard and shared memories of recent events seem to be indistin-
guishable from the narrator's own impressions of the disaster. By bringing
together fragments of the past, he hopes to rescue those 'minor' voices that
might otherwise be silenced in more conventionally structured narratives. In a
letter to his editor that accompanies the narrative, Saint-Éloi explains the aims
and motivations behind the aesthetics of his account:

> I wrote this book to accompany the lullaby lament of *goudou-goudou* deep-
> rooted in all Haitians […] I think about my mother, who collects different
> stories so as to create her own story. I did something similar by merging all
> those stories of the earthquake so that nothing was forgotten. Eventually, I
> do not know anymore which story belongs to me and which to others.[6]

Saint-Éloi sees his literary pursuit, in other words, as an attempt to interweave
and represent the multitude of diverse experiences that the earthquake elicits.
Narrativisation is not synonymous with recording the unfolding of events; rather,
it signifies the possibility of a creative interweaving of first-hand experience with
other survivor accounts. This act of incorporation and reconstruction becomes an
ethical gesture of opposition against those kinds of silencing narratives that are
dominated by just one voice or just one author. In effect, Saint-Éloi's text is no

longer 'authored' at all but is rather reassembled by a narrator-figure who is not directly synonymous with the author. By destabilising the idea of singular authorship and fixed narratorial subjectivity, the narrator implicitly undermines any claims to the historical authority of his text. Rather, *Haïti, kenbe la!* should be treated as one of many possible sources of knowledge of the 2010 disaster and not as a definitive record of the event or a definite and unchangeable 'past'.

The account's formal features, when coupled with its thematic concerns and the self-reflexive character of the narrative, mark out *Haïti, kenbe la!* as a complex artistic work that opposes easy categorisations of the event as either disconnected from the past or just as another of Haiti's many failures. It also suggests new, future-oriented modes of narrative engagement that go beyond traditional means of literary expression and conventional narrative form. Any subsequent portrayals of the event must therefore engage with the disaster's wider past and the discursive representations of Haiti in which it was framed. *Haïti, kenbe la!* accordingly embeds the earthquake within the long histories of imperialism and colonialism, depicting it as one of the many metaphorical 'tremors' that have shaken the country, but also as the potential beginning of a new, revisionist history. For the narrator, the foundational 'earthquake' in the country's history happened in 1492, with Columbus's 'discovery' being the first in a series of massive upheavals that punctuated the nation's violent past; for him '[t]he first earthquake in our history is called Christopher Columbus'.[7] In an extension of the metaphor, the history of the country is envisioned, both by Saint-Éloi and Oliver-Smith (Part I), as a sequence of tremors followed by regular aftershocks – a potent image also invoked in anthropological analyses of 12 January as 'the culmination of [Haiti's] own more than 500-year earthquake'.[8] Paraphrasing the words of his grandmother Grann Tida, the narrator asserts: 'The country's history since its independence is a sequence of earthquakes followed by regular aftershocks.'[9] In this reformulation, the narrator plays with the contrasting ideas of a singular historical event and a repeated one. Despite the potential of the metaphor to emphasise long-term processes and violences that shaped the country's past and present, Saint-Éloi does not want to suggest, however, an unchanging and circular vision of the country's past or its future. 'Réplique', 'aftershocks', in its wider sense is also a reply and a retort, an objection,[10] not just a copy or a duplicate of an original object or event, and implies a response as well as the taking of a stance. In effect, in its layered meaning as an aftershock but also as a reply and an objection, the term points to the agency and resolve that equally shapes the country's longer history.

At the same time, the contrast between repetition and singularity asserts that this particular disaster, in its scale and impact, is indeed an unprecedented occurrence yet still one that does not exist outside of previous global historical and political processes. These experiences, transmitted to him through collective memory and family stories, 'seem to constitute memories in their own right',[11] and to contribute to the polyphonic quality of his account, with these earlier traumatic events, recalled as 'postmemory',[12] reverberating in this composite narrative of the most recent disaster. Yet, the echoing roar of *goudougoudou* does not strike the last chord in Saint-Éloi's account nor in Haiti's history. Like the reverberations of the original colonial violence and Haiti's victory over it, the tremors can become, as the narrator hopes, just one among many melodic lines that come together in a defiant polyphony of the country's recent past.

This metaphorical reimagining of history is, at the same time, a gesture of critique directed against popularised depictions of Haiti's past and present predicament as a paradoxical repetition of 'unique' events, 'beyond comparison' – common representations that depoliticise and marginalise the country. Several years after the 2010 earthquake and despite growing critical scholarship, attention-grabbing newspaper titles and media coverage – for example 'Beauty vs. Poverty'[13] and 'The graveyard of hope'[14] – continue to reaffirm homogenising images of post-earthquake and, more recently, post-Matthew Haiti as 'discarded', extreme space, and Haitians as 'devastated' people. History and myth collide in those popular portrayals, where Haiti is presented as at once disaster-*prone* and disaster *zone*. Such representations configure temporality in triadic terms: the country's past is described as 'heroic'; its present is conceived in terms of the 'affirmation of resilience' and the future is phrased in terms of 'unfulfilled hopes' and 'lack of change'.

Over and against such antithetical and extreme formulations, the narrator of *Haïti, kenbe la!* examines the ideological nature of history and historical narratives, which far from being 'neutral' categories are discursive constructs in which objectified images of Haiti and its people continue to be manipulated in both 'epic' and 'quotidian' ways:

> The cycle of national history varied, depending on the author, between the grandeur of an epic dream and everyday depravation. The image that remains of us: the seriousness of a nation at once crazy and heroic. Either we laugh or we cry. Never anything in between or in moderation.[15]

By juxtaposing these contrasting yet equally objectifying qualifiers, the narrator establishes a direct correlation between two misrepresentations of the country's past. His narrative intervention thus becomes a creative attempt to give a more nuanced view of the ways in which Haiti's pasts and futures can be remembered and imagined – a view that destabilises the discursive and epistemological constructs within which the country was previously imprisoned.

In order to make audible these new, empowering narratives of Haiti's recent history, the narrator elevates the importance of memory and complex layered memories over seemingly univocal archival records: History (with a capital 'H') is, he suggests, 'a memory of the times' – 'la mémoire du temps'[16] – not just a simple record of what happened. As he says: 'History is a memory of the times. It is the part of us we always dream of.'[17] This new History he hopes to re-collect and write down is very much a collective undertaking, a 'work-in-progress' that seeks to remember, recall, but also to undo, previous histories and archival accounts that reflected imperial power structures: 'The winners will always own the archives. They have their faded books and versions of all histories.'[18] Rather than being an objective and comprehensive source of knowledge, the historical record, as suggested metaphorically here by Saint-Éloi and earlier argued by historians Jacques Le Goff and Michel-Rolph Trouillot, 'expresses past society's power over memory and over the future: the document is what remains'.[19] These archival artefacts of the past in turn create an exclusive and silencing account which, in the postcolonial context of Haiti, becomes a reiteration of the colonial and neocolonial discourses to which Haiti has been subjected over centuries and in which the country's voice is, in the extreme mode, either a hushed lament over one's misfortune or a desperate call for help, never a reasoned and deliberate expression of a people.

Echoing some of the scholarly debates on the non-neutral character of historical narratives and the power dynamics involved in creating a historical archive,[20] the narrator's definition of History as 'a memory of the times' accentuates the open-ended and subjective character of all historical accounts. This reformulation, however, is not without its problems. At times, the narrator positions memory as an antithesis of history, while at others he seems to equate memory with history, a tension that, in Le Goff's gloss, 'seem[s] virtually to identify history with memory, and even [gives] preference in some sense to memory, on the ground that it is more authentic, "truer" than history, which is presumed to be artificial and, above all, manipulative of memory'.[21] Yet as Le Goff makes clear, the two are neither mutually exclusive nor simply synonymous but are best conceived as interdependent and complementary threads:

> Memory is the raw material of history. Whether mental, oral, or written, it is the living source from which historians draw. [...] Moreover, the discipline of history nourishes memory in turn, and enters into the great dialectical process of memory and forgetting experienced by individuals and societies.[22]

For Le Goff, both memory and history are dynamic and fluid categories, and are continually subject to revision and change. They co-exist in a dialectical relationship: they are different forms of remembering that bleed into each other. Aware of possible manipulations of history, he emphasises the continuing need to re-assess our knowledge of the past and to question the sources and forms of this knowledge – a call taken up creatively by Saint-Éloi.

Sharing some of Le Goff's preoccupations, Trouillot translates them into more specifically postcolonial contexts, pointing to the ways in which the production of history is a dynamic, inherently ambiguous process and a site of a power-play:

> For what history is changes with time and place or, better said, history reveals itself only through the production of specific narratives. What matters most are the process and conditions of production of such narratives. Only a focus on that process can uncover the ways in which the two

sides of historicity intertwine in a particular context. Only through that overlap can we discover the differential exercise of power that makes some narratives possible and silences others.[23]

For Trouillot, history is a discursive site of continued negotiation and confrontation between multiple narratives and voices, some of them deliberately silenced at various points in time. However, rather than valuing some narratives of past events over others, as Anderson's piece (see Part I) seems to do in relation to post-disaster cultural responses, Trouillot stresses the importance of critically comparing 'the two sides of historicity [in each] particular context'.[24] The narrator's distinction between the 'memory of the times' and the 'record of times' in *Haïti, kenbe la!* in many ways echoes Trouillot's methodological scrutiny and acts as a meta-thematic comment. Saint-Éloi is aware that the narrative he is creating also has a historical dimension and that narrativisation is in itself a process of selection and editing. Yet by interrogating his own narrative practice and redefining it as a process of recalling, of remembering and undoing the past simultaneously, he hopes that his account can provide a heteroglossic presentation of the earthquake and its aftermath that is empowering for all Haitians, both now and in times to come.

Towards a new history

This critical awareness of the tensions and ambiguities inherent in the concepts of 'the past' and 'history' does not entail, however, 'an absence of purpose'[25] or the impossibility of taking an ethical position. On the contrary, it can determine one's engagement in current socio-political struggles, and it is precisely this realisation of the exclusive character of narrative accounts of the past, and of their continuing disempowering impact, that animates *Haïti, kenbe la!*'s insistence on writing 'new History' in the wake of the disaster. The creative process in turn acquires a trans-temporal character: it is an act of historical revision that partakes in contemporary struggles. In this context, as Trouillot posits and Saint-Éloi creatively enacts, the rewriting of collective history is itself a mode of action:

> But we may want to keep in mind that deeds and words are not as distinguishable as we often presume. History does not belong only to its narrators, professional or amateur. While some of us debate what history is or was, others take it in their own hands.[26]

Against the scholarly tendency to compartmentalise history as a distinct entity available for scrutiny only to professional historians – a position that potentially masks detachment from current political concerns and crises – Trouillot suggests a view of history as open-ended and collectively created and owned.[27] By using the idiomatic expression of 'taking something into one's own hand', he pairs the idea of direct action with the re-envisaging of history as a discursive category.

Saint-Éloi responds to this call for a dual reformulation of history in a writerly manner by taking up his pen and initiating, at least on the pages of his account, a process of reworking traditional narrative structures, accompanied by a self-reflexive investigation into the nature of history in the aftermath of the earthquake, with the two-fold attempt acquiring a direct ethical significance.

Here, the complex texture of *Haïti, kenbe la!* is in itself an attempt to inaugurate and embody this more polyphonic account, Haiti's 'new History', which would make audible both the silenced voices of the past and those muted by the recent disaster: 'The country's new history starts with the piercing cry which cracks the centre of the earth: the gust of machine guns, the shaking of the roofs, the cracking of chairs'.[28] The highly emotive and onomatopoeic language of the passage mimics the ravage caused by the earthquake – the sudden strike of *goudougoudou*. It also echoes an earlier, symbolic and highly evocative, formulation of the beginnings of Creole literature, inaugurated in '[a] sudden *cry* arising from the ship's hold' [italics original].[29] In joining again the images of the first colonial tremors with those scenes of post-earthquake destruction, the narrator asserts that also this more recent roar cannot and must not be silenced, marking a second point of temporal rupture and the violent beginning of a 'new History'. In effect, *Haïti, kenbe la!*'s poetically charged formulation of this time frame is intricately linked to the ethical aims that underpin its aesthetic concerns: the creative reshaping of the past as a dual category, characterised from now on by its untimeliness, becomes an attempt to formulate new discourses on Haiti and a new ethics of collective remembrance in the disaster's wake.

A similar pairing of literary aesthetics and historical enquiry can be found in Walter Benjamin's work whose critical writings on history and progress help to reveal the complexity of *Haïti, kenbe la!* and point to its direct ethical and political engagement. The philosopher, in metaphorically charged terms, opposes the view of history as progress, whether this is understood within the framework of bourgeois or orthodox Marxist ideology, defining it, instead, as a form of ethical remembrance that violently disrupts any sense of continuity. Merging theological and political discourse, Benjamin sees this violent break as 'the only point of contact between a postlapsarian world and a redeemed world',[30] which occurs 'when the Messiah or the messianic moment interrupts the continuity of history and brings history itself to a halt'.[31] As such, history, in Benjamin's radical reconfiguration of the term, 'is not simply a science but also a form of remembrance, '*Eindenken*'[32] that, in Rebecca Comay's gloss, 'denotes the moment of '"blasting open" the continuum'[33] and announces 'a mindfulness or vigilance which refuses to take in (or to be taken in by) a tradition authorizing itself as the continuity of an essential legacy, task or mission to be transmitted, developed or enacted'.[34] Consequently, this view of history, which is inspired by the Judaic Messianic tradition[35] and speaks of 'the necessity to lift a historical period or event out of the continuum',[36] aims to redefine the existing 'formulations of the concepts of history and historical origin'[37] – an ambition clearly visible in Saint-Éloi's narrative.

In effect, away from their presumed linearity, history and progress, in Benjamin's reformulation, are no longer viewed as teleological. Rather, '[t]he concept of progress must be grounded in the idea of catastrophe',[38] and 'has its seat not in the continuity of elapsing time but in its interferences – where the truly new makes itself felt for the first time, with the sobriety of dawn'.[39] Instead of conceiving of temporality in terms of clear divisions between past and present, Benjamin, and similarly to him, Saint-Éloi, ask us then to think about historical time as a relationship between the 'now' and the 'then'. These concepts are brought together, for the philosopher, in the historical object and the figure of the historian as collector: 'It is owing to this monadological structure that the historical object finds represented in its interior its own fore-history and after-history'.[40] As a form of remembrance, history is characterised by a sense of incompleteness, and this recognition enables Benjamin to collect a narrative that restores justice to the vanquished who 'may have lost the war, but [...] did not lose their history'.[41]

Embodying his critical methodology in the form of literary montage, Benjamin compiles a narrative collage of seemingly unimportant and unrelated objects. He opposes the idea of a linear record of history, challenges the arbitrary valuation of one historical object over the other, and rescues these discarded objects from the refuse of history. Benjamin's method resonates in Saint-Éloi's mosaic account, which features multiple voices and seemingly disjointed memories as part of the narrator's attempt 'to perform a small but never insignificant *restitutio in integrum*,[42] if only by doing as little as simply refusing to accept the finality of past suffering'.[43] Much like Benjamin, Saint-Éloi underscores the urgency of remembrance, stressing the responsibility that lies with survivors who, in preserving the memory of the vanquished, can subvert dominant historical discourses as well as their recent revisions and create new history as a polyphonic 'memory of the times'.

Yet this restoration of justice, as envisaged and performed creatively in *Haïti, kenbe la!*, and one which is rooted in the experience of the catastrophe, challenges and goes beyond the legal category of *restitutio in integrum*. Justice is not simply 'reversing history', going back to the 'previous condition'. *Haïti, kenba la!* calls for a different type of restitution and justice by refusing, with dignity, to accept the finality of past suffering – the disaster is not yet over – and by making clear that no return to the pre-earthquake conditions is possible, nor should it be wished for. *Restitutio in integrum*, if indeed the restoration is to be complete, entails the elimination of the structures of exploitation and the discourses of Haiti's 'permanent' vulnerability that is further increased in the post-disaster context of the cholera epidemic, for example, through denial and neglect. Pointing to this importance of struggle for presence and recognition, against such discourses and practices of erasure, one of the narrator's friends calls out: 'We need to get up, stand on our own feet, my friend. With or without the earthquake.'[44] His request is taken up in the book's title, *kenbe la!*, hold on there, remain. An extended response to this call, the process of constructing a new narrative of the country's recent history has, then, a clear ethical dimension: by refusing to recover previous linear and seemingly fixed divisions of time and discourses of the country's past, Saint-Éloi's account of the event allows for a restoration of collective justice, albeit of a limited kind.

Reimaging the future

The 2010 Haitian earthquake, as I have been suggesting here via the works of Martin and Saint-Éloi, has provided a moment of redefinition of the disaster as a closed-off occurrence and of the past as a discursive category and historical construct. It has inaugurated a 'new history', both personal and collective, which has had a defining impact on the ways in which the immediate and long-term future can be understood and envisaged beyond the sense of a simple repetition or prolongation of a past event. Martin's narrative accentuates the uncertainties attendant on this, the extreme difficulty of predicting and preparing for a future experienced as a state of suspense in which the possibility always exists of further devastating aftershocks. The question the narrative asks is how to think of the future where recurring environmental hazards – such as seasonal tropical storms, or tremors – are at once inevitable and unpredictable, halted yet only temporarily. Still, however painful this might be, haunting impressions of the disaster must not be effaced if the memory of those lost under the rubble is to be preserved. Saint-Éloi's account shares this anxiety, but also gestures towards a more hopeful beginning – one rooted in revised narratives of the past – even though the larger history within which this is contained can at times seem to oscillate between violent extremes.

In *Le Tremblement*, the narrator's personal experience of the future is characterised by a sense of temporal uncertainty and a possible repetition of the initial event. Echoing, to some extent, the tone of Samuel Beckett's absurdist drama *Waiting for Godot* (1949), *Le Tremblement* encodes the waiting experience in negative terms of inertia and disempowering passivity. Immediately after the earthquake, the narrator feels caught within a seemingly interminable state of stasis. The repetition of the words 'waiting', 'immobility', 'suspended time'[45] throughout the narrative, underscores an overwhelming sense of entrapment from which there appears to be no escape. The narrator's role is reduced to the plain act of waiting: 'there is nothing to do but wait. And the time passes.'[46] His vision of the future anxiously combines his legitimate belief that aftershocks will follow in the wake of the original tremors with his total inability to predict when these might occur. As the afternoon of 13 January passes, and the fatal hour again draws near, the narrator finds himself counting down the time to possible aftershocks:

> It's three in the afternoon. We are waiting for a strong aftershock. [...] Can we trust the history of the last twenty-four hours? Will the earth shake again precisely at 4.52 p.m.? Or is it only a myth supported by empty statistics?[47]

Trapped in a state of paralysis, he can neither trust the past nor the future since these temporal categories either revert to the initial experience of the disaster or signify a repetition of it. Extreme anxiety is the result:

[...] the earth, so tranquil, innocent and child-like at the moment, will tremble again; 'the aftershocks' will surely follow because it is a rule that a strong tremor is followed by a sequence of shocks, which are usually smaller, erratic and can spread over days so that we cannot predict when they occur. We're caught in this improbable expectation.[48]

If scientific knowledge at least provides a simulacrum of certainty, the narrator cannot imagine the future any differently than as a mode of waiting for an unavoidable repetition of the initial event. Waiting is both anticipation of change and, paradoxically, an act of deferral of the future. It is a dialectical state directed in its positive modality 'toward something that is desired',[49] and in its negative modality 'toward something that is not desired: it is dread'.[50]

Distinguishing between waiting for *something* and waiting for *anything*, the anthropologist Vincent Crapanzano suggests that waiting is a predominantly passive state which produces 'feelings of powerlessness, helplessness, and vulnerability'.[51] It is organised around the principles of '[the] backward glance, [the] seeking [of] security in the experience of the past, [the] taking solace in history, [all of] which devitaliz[e] the here and now'.[52] For the narrator of *Le Tremblement*, though, even the past can offer no sense of security as previous knowledge and possibilities of reaction seem non-transferable to the new reality of the disaster. The rhythms of seasonal change are no longer a source of reassurance. Instead, after the earthquake, weather assumes a distinctly threatening dimension, signalling the potential repetition of the disaster:

> We would be almost well in our untroubled, carefree vision of the future because the future in the Greater and Lesser Antilles is always the same, the days follow one another without the interference of the seasons [...] except when a hurricane passes or a volcano roars like a cow with swollen udders [...] except when the earth shakes.[53]

In the Haitian context, the rhythmic recurrence of seasonal change cannot be separated from the equally regular, yet still inherently erratic, environmental risks and hazards that, with Haiti's increasing vulnerability, threaten to become seasonal disasters. In this sense, to wait for the change is to experience waiting in its negative modality: as dread.

Similarly, in *Haïti, kenbe la!* the immediate future, both on a personal and community level, has a two-fold character, being envisaged in terms of the possibility of a new, hopeful beginning, but also as a heightened consciousness of the devastating scale of destruction and loss. The two merge in the apocalyptic rhetoric that surrounds the account's vision of the future:

> The first morning resembles the beginning of the world. Since the morning of the 13th of January the tennis courts have been gradually emptying. Distraught, some cling onto a web of illusions. Paralysed by the sun, they hang about helpless near the swimming pool. Others expend incredible energy to deny defeat. They move and refuse to admit that the earth has trembled.[54]

The first day of this new world is a time during which the survivors struggle to affirm their very existence. Described as wild, distraught, blinded by the light, they are helpless in the face of the destruction that visibly surrounds them. Saint-Éloi's densely metaphorical language conveys the double – psycho-logical and physical – impact of the event on those it has subjected to its whim. Yet this apocalyptic vision is later juxtaposed with a more positive representa-tion in which the extraordinary resolve of the survivors is underlined: 'The calm, unshakeable sun extends already over the city. Wednesday, the 13th of January. 8.15 am. It's a clear morning full of promises. Everyone is looking for a reason to defy despair.'[55] In this later reflection on the near-future, the sun is no longer depicted as bullying, but is rendered as calm and unshakeable. It is a clear morning, 'bursting with promises' ('gorgé de promesses'), and people are looking for a way to abandon the otherwise overpowering feeling of despair.

This affirmation of sense of purpose proves only momentary, however, and the tension between defiance and denial permeates even the most positive depictions of 13 January. The narrator's emphasis on the survivors' struggle, manifested in the repetition of everyday gestures, suggests their refusal to acknowledge the immensity of the disaster. By suppressing the unprecedented impact of the event, they can create a sense of relative continuity and control, though this also turns out to be illusory. Analysing the reactions of some of his fellow survivors, the narrator remarks: 'Some negate the scale of the earthquake repeating everyday gestures with the same will.'[56] The recourse to daily habits is an attempt to assert some degree of agency, to defy the scale of the disaster and to counteract the impact of the experience on one's sense of self. Yet in the face of such staggering loss, these actions cannot be sustained for long and the effort to carry on with everyday existence soon gives way to disillusionment and desperation. This shift in perception is reflected in the narrator's revised description of the day after: 'The day after is even worse. We discovered that nothing makes sense anymore. Even assurances that the earth is solid.'[57] Here, the day after the earthquake is seen as being worse than the day itself, since it forces the narrator and other survivors to face the inadequacy of finding words to express what has happened to them and to confront 'the familiarity of despair'.[58]

The future the narrator now envisages is one that continues to be defined by the untimely presence of the event. Having acknowledged the full force of the disaster, he still struggles to assess its scale. The narrator trawls through various literary representations of post-apocalyptic landscapes, frenetically seeking com-parisons for the destruction he has just witnessed:

> I ask myself in which novels and films I've already seen those vanishing sil-houettes. Those cities dismantled by the catastrophe. I close my eyes and remind myself of specific scenes in *Blindness* by José Saramago, *The New York Trilogy* by Paul Auster, *The Road* by Cormac McCarthy. Disaster. Desolation. Paradise lost.[59]

By attempting to establish analogies between the real ravaging of the city and familiar symbolic images of desolation and debris, the narrator hopes to contain the scale of the disaster within known narrative frameworks. But no image of fictional ruination, however familiar, can help him come to terms with the real havoc he has encountered. In a moment of sobering realisation, he admits: 'These landscapes to which I belong are not fictional scenes.'[60]

Moreover, the sheer scale of devastation and loss also determines the ways in which *Haïti, kenbe la!* envisages the more distant future in the aftermath of the earthquake. The passing of time can bring a difficult-to-accept yet still-longed-for confirmation (e.g. of death), or it can turn into a prolonged experience of anticipation. Painful though it is, for Saint-Éloi as for Martin, only the certainty of death can allow the narrators to move on. At the same time, the impossibility of mourning and achieving a sense of closure severely limits their ability to envisage the future other than as a mode of impotent waiting as remarked by Saint-Éloi:

> How to mourn a loved one if we're not sure if he or she is dead? [...]. How to cry when we're secretly waiting for the return of someone who has left in the morning? We wait for the person on the evening of the 12th of January and the days to come, in each rustle of the feet, in each aftershock. We risk waiting *ad vitam.*[61]

In passages like this one, the inevitable passing of time is contrasted with the experience of temporality as a state of interminable *attente*, a life on hold. Seen in these terms, the future, on the personal level and for those directly affected, far from signifying a new beginning, becomes, on the one hand, a laboursome and defiant wake and, on the other, possibly life-long prolongation of the original event that shapes the untimely interventions these texts make.[62]

Conclusion

'On Tuesday, January 12, 2010, eternity lasted less than sixty seconds [...] and forever altered the landscape of a city, a country, and our memory.'[63] However it is framed, the sudden and arbitrary character of the experience, which is affirmed in Evelyne Trouillot's remark and explored in detail in *Le Tremblement* and *Haïti, kenbe la!*, contrasts with its lasting personal and collective impact. The earthquake is both a point of rupture and an untimely unfolding event that disrupts conventional notions of time as based on the separation of past, present and future. In neither account can the process of narrativisation and its formal framings, or the act of distancing oneself from the site of the disaster, lessen the event's continuous impact. In both texts, instead, the untimely character of *goudougoudou* is made manifest in the disruption of previous temporal categories. The earthquake is a punctual event that is presented initially as a defined, measurable occurrence in the past, but then becomes a radical point of rupture marking a separation between the new temporal divisions of 'before' and 'after'. A break in its own right, the earth-quake, however, is equally embedded within national and global histories that, jointly, determine the force of its blow.

In so far as they both emphasise the personal, collective and historical significance of the 2010 tremors, *Le Tremblement* and *Haïti, kenbe la!* portray the earthquake in terms of the beginning of a new history. Fully aware of their duty to preserve the painful memory of the event, they accept, in the words of the anthropologist and writer Laura Rose Wagner, that although '[t]he initial moment of rupture is over [...] this disaster has no foreseeable end'.[64] This lack of closure is equally a refusal to mark as 'finished' and as a closed-off 'past', effectively muting and erasing the suffering of those who survived and who died in the earthquake. In Martin's prayer-like memoir, this gesture involves the detachment from previous personal history and the wholesale redefinition of earlier experiences and memories of Haiti in light of the event. Present and future merge into one in an experience of temporal suspense in which the seemingly unavoidable reprise of *goudougoudou* is fearfully antici-pated. In this context, the future, at least for now, becomes but a mode of repe-tition of the past, and is filled with echoes of destruction. For Saint-Éloi, meanwhile, the earthquake challenges the meanings of history itself while also offering an opportunity – however painful it might be – to recall earlier narrat-ives of national history and to create an empowering, ethically committed account of History as the memory of the times. The new temporal category the event inaugurates is at once a hopeful and a frightening one, necessitating the acceptance of loss and the confrontation of mass destruction. Yet this recogni-tion is not an entirely immobilising one. On the contrary, it motivates the nar-rator to challenge the idea of the past as a stable and neutral category, offering a polyphonic and open-ended account of the multifaceted experience of the disaster.

Varying though they are in their narrative form, in their ways of framing the event, and in the provisional conclusions they offer for interpreting its collective significance, *Le Tremblement* and *Haïti, kenbe la!* are both powerfully untimely narratives of individual and collective survivorhood. Against a historical back-ground of disempowering and silencing narratives, such narratives forge a new ethics of writing: a dialogue addressed to and commemorating the lost voices of the disaster. The force of the tremors permanently reshapes Haiti's rural and urban landscapes, and the jolts' untimely reverberations forever disrupt indi-vidual and collective definitions of temporality and history, challenging, in effect, event- and site-bound notions of disaster, ruin and ruination – key cat-egories of time and space to which I now turn. Urban locations, rural sites or the all-surrounding rubble, all have their distinct yet intertwined histories that are taken up, exposed or negated in the oft contrasting visions of post-quake recon-struction, rebuilding and remaking.

Notes

1 '[…] il n'y a aura plus jamais de « comme avant », et que la rupture, même colmatée, comblée de plâtre, de mortier, restera vive, à vif dans les murs et dans l'homme mûr que, désormais, définitivement je suis, sans espoir de retour'. Martin, p. 49.

2 'Nous, quoi que nous laissions derrière nous, c'est un peu de nous-mêmes que nous nous délesterons, d'un peu de notre histoire – pour, il est vrai, si nous nous sortons de cette débâcle, rejaillir dans une nouvelle histoire.' Martin, p. 81.

3 'délestrer', in *Reverso: Collins Online English-French Dictionary* (HarperCollins Publishers, 2012) http://dictionary.reverso.net/french-english/d%C3%A9lestrer [accessed 10 April 2013].

4 Saint-Éloi, p. 19.

5 'Je dois désormais compter avec les victimes et ce *godou-godou* qui a chambardé une certaine historie du pays. On vivra *après* toute sa vie pour classer les choses en deux temps: *avant* le séisme et *après* le séisme.' [italics added].

> I need to include (myself) among the victims of the *goudou-goudou* which has turned upside down a certain history of the country. We will all live our whole lives classifying things into two tenses: before the earthquake and after the earthquake.
>
> Saint-Éloi, p. 38

6 J'ai écrit ce livre pour accompagner d'une berceuse ce cri *goudou-goudou* enraciné dans les entrailles de tous les Haïtiens. […] Je pense à ma mère qui recueille des histories pour inventer son propre récit. Je fais pareil, en enfonçant dans ma tête toutes ces histoires de séisme pour que rien ne soit oubli. À la fin, je ne sais plus ce qui tient de moi ou des autres.

> Saint-Éloi, p. 266

The use of the future tense 'on vivra', which signifies both 'we will live' and 'one will live', underlines the ways in which the event strengthens the relationship between the narrator and his fellow survivors. Immersed together in this state of shared victimhood, they are aware that in the aftermath of the disaster, previous temporal divisions are no longer valid. The past, no matter how remote, is now synonymous with 'before the earthquake', and present and future merge into one composite temporal category: 'after the disaster'.

7 '[l]e premier séisme de notre histoire s'appelle Christophe Colomb'. Saint-Éloi, p. 202.

8 Oliver-Smith, in *Tectonic Shifts*, pp. 18–23 (p. 18).

9 'l'histoire du pays depuis l'indépendance est une suite de séismes suivis de répliques régulières'. Saint-Éloi, p. 215.

10 'réplique', in *Reverso: Collins Online English-French Dictionary* (HarperCollins Publishers, 2012) http://dictionary.reverso.net/french-english/replique [accessed 1 June 2017].

11 Marianne Hirsch, 'The Generation of Postmemory', *Poetics Today*, 29 (2008), 102–28 (p. 107).

12 Hirsch uses the term 'postmemory' to describe' the relationship of the second generation to powerful, often traumatic, experiences that preceded their births but that were nevertheless transmitted to them so deeply as to constitute memories in their own right. Here, I use the term in a wider sense to illustrate this passing on of foundational traumatic memories around which individual identity is established. Hirsch, p. 103.

13 Trenton Daniel, 'Beauty vs Poverty: Haitian Slums Get Psychedelic Make-Over in Honour of Artist Prefete Duffaut's *Cities in the Sky*', *Independent*, 26 March 2013 www.independent.co.uk/arts-entertainment/art/news/beauty-vs-poverty-haitian-slums-get-psychedelic-makeover-in-honour-of-artist-prefete-duffauts-cities-in-the-sky-8549 331.html [accessed 02 May 2013].

14 Peter Popham, 'Haiti: The Graveyard of Hope', *Independent*, 13 January 2013 www.
    independent.co.uk/news/world/americas/haiti-the-graveyard-of-hope-8449775.html
    [accessed 02 May 2013].

15   Le cycle de l'histoire nationale oscillait, d'après ceux qui l'écrivaient, entre la
        grandeur d'un rêve épique et la dépravation du quotidien. L'image qui reste de
        nous: la gravité d'un peuple à la fois héroïque et loufoque. Soit on rit. Soit on
        pleure. Jamais entre le deux. Jamais la mesure.

                                                                    Saint-Éloi, p. 203
16 Saint-Éloi, p. 198.

17 'L'Histoire, c'est la mémoire du temps. Cette part de nous, toujours rêvée'. Saint-
    Éloi, p. 198.

18 'Les vainqueurs ont toujours pour eux les archives. Ils ont leur livre jauni et les ver-
    sions de toutes les histoires.' Saint-Éloi, p. 199.

19 Jacques Le Goff, *History and Memory*, trans. by Steven Rendall and Elizabeth
    Claman (New York, NY: Columbia University Press, 1992), p. xvii.

20 Two key voices in this debate – Jacques Le Goff and Michel Trouillot – have ques-
    tioned conventional readings of history and the sources used to construct it, seeing
    such accounts as insufficiently self-reflexive. The French historian (Le Goff) and the
    Haitian anthropologist (Trouillot) both challenge the seemingly objective character of
    archival documents, highlighting instead the power dynamics involved in the creation
    of historical narratives.

21 Le Goff, p. xi.

22 Le Goff, p. xi.

23 Michel-Rolph Trouillot, *Silencing the Past: Power and the Production of History*
    (Boston, MA: Beacon Press, 1995), p. 25.

24 Trouillot, p. 25.

25 Trouillot, p. 153.

26 Trouillot, pp. 152–3.

27   The more historians wrote about the past worlds, the more The Past became real
        as a separate world. But as various crises of our times impinge upon identities
        thought to be long established or silent, we move closer to an era when profes-
        sional historians will have to position themselves more clearly within the present,
        lest politicians, magnates, or ethnic leaders alone write history for them.

                                                                    Trouillot, p. 152

28 '[l]a nouvelle histoire du pays débute par ce cri perçant qui fendille le ventre de la
    terre: rafales des mitrailleuses lourdes, tremblement des toits, craquelure des chaises'.
    Saint-Éloi, p. 178.

29 '*un cri* surgissant de la cale. Celui d'un Africain quelconque.' Patrick Chamoiseau,
    Raphaël Confiant, *Lettres créoles: Tracées antillaises et continentales de la littérature
    Haïti, Guadeloupe, Martinique, Guyane 1635–1975* (Paris: Hatier, 1991), p. 32.

30 Bram Mertens, '"Hope, Yes, But Not For Us": Messianism and Redemption in the
    Work of Walter Benjamin', in *Messianism, Apocalypse and Redemption in 20th-Cen-
    tury German Thought*, ed. by Wayne Cristaudo and Wendy Baker (Adelaide: ATF
    Press, 2006), pp. 63–79 (p. 72).

31 Mertens, p. 72.

32 Benjamin, p. 471.

33 Rebecca Comay, 'Benjamin's Endgame', in *Walter Benjamin's Philosophy: Destruc-
    tion and Experience*, ed. by Andrew Benjamin and Peter Osborne (London and New
    York, NY: Routledge, 1994), pp. 251–92 (p. 266).

34 Comay, p. 266.

35 The term messianism is derived from messiah, a transliteration of the Hebrew mashiaḥ ('anointed'), which originally denoted a king whose reign was consecrated by a rite of anointment with oil. In the Hebrew scriptures (Old Testament), mashiaḥ is always used in reference to the actual king of Israel: Saul (1 Sm. 12:3–5, 24:7–11), David (2 Sm. 19:21–22), Solomon (2 Chr. 6:42), or the king in general (Ps. 2:2, 18:50, 20:6, 28:8, 84:9, 89:38, 89:51, 132:17). In the intertestamental period, however, the term was applied to the future king, who was expected to restore the kingdom of Israel and save the people from all evil.

    Helmer Ringgren, 'Messianism: An Overview', in *Encyclopedia of Religion*, ed. by Lindsay Jones (Detroit, MI: Macmillan Reference, 2005), pp. 5972–4 (p. 5972)

36 Mertens, p. 72.

37 Peter Osborne and Matthew Charles, 'Walter Benjamin', in *The Stanford Encyclopedia of Philosophy Online*, ed.by Edward N. Zalta (2012) http://plato.stanford.edu/archives/win2012/entries/benjamin/ [accessed 20 April 2013].

38 Walter Benjamin, 'N: On the Theory of Knowledge, Theory of Progress', in *The Arcades Project*, ed. by Rolf Tiedemann, trans. by Howard Eiland and Kevin McLaughlin (Cambridge, MA. and London: The Belknap Press of Harvard University Press, 1999), pp. 456–88 (p. 473).

39 Benjamin, p. 474.

40 Benjamin, p. 475.

41 Mertens, p. 75.

42 The *Oxford Dictionary of Law Online* defines the term simply as 'Restoration to the original position'. Jonathan Law and Elizabeth A. Martin, 'restitutio in integrum', in *A Dictionary of Law Online* (Oxford University Press, 2009) www.oxfordreference.com/10.1093/acref/9780199551248.001.0001/acref-9780199551248-e-3401 [accessed 24 February 2015]. *OED* adds to the definition: 'Placement of an injured party in the situation which would have prevailed had no injury been sustained; restoration of the status quo ante.' 'restitutio in integrum, n.', in *Oxford English Dictionary Online* (Oxford University Press, 2015) www.oed.com/view/Entry/163965?redirectedFrom=restitutio+in+integrum [accessed 24 February 2015]. Finally, the *Guide to Latin in International Law Online* provides a more contextualised definition of the term:

> Restitution of a damaged or taken thing to its previous condition, as by the restoration of the thing (or a comparable substitute) to the owner who has been deprived of it. In modern times, the term has sometimes been used to refer to the payment of full compensation for the loss sustained in the event that restoration or replacement of the thing damaged or taken is not feasible (for example, when it has been depleted or destroyed).
>
> Aaron X. Fellmeth and Maurice Horwitz, 'restitutio in integrum', in *Guide to Latin in International Law Online* (Oxford University Press, 2009) www.oxfordreference.com/view/10.1093/acref/9780195369380.001.0001/acref-9780195369380-e-1856 [accessed 24 February 2015]

43 Mertens, p. 75.

44 'On doit être debout, mon cher. Séisme ou pas.' Saint-Éloi, p. 220.

45 'attente', 'immobilité, 'temps suspendu'.

46 'toujours il n'y rien à faire – qu'attendre. Et le temps passe.' Martin, p. 82.

47 'Il est quinze heures. Nous attendons la forte réplique [...] Peut-on se fier à cette histoire des vingt-quatre heures? La terre va-t-elle de nouveau trembler à précisément 16h 52? Ou n'est-ce qu'un mythe étayé de vaines statistiques?' Martin, p. 69.

48 [...] la terre, cette terre pour l'heure si quiète, innocente, enfantine, elle allait de nouveau trembler; des « répliques » allaient nécessairement survenir, parce que c'est la règle et qu'à toute forte secousse font suite des tremblements de normalement

moindre importance, mais erratiques et devant s'étaler sur des jours et des jours, sans qu'on puisse en rien les anticiper. Nous attendons dans cette attente improbable.

Martin, p. 22

49 Vincent Crapanzano, *Waiting: The Whites of South Africa* (New York, NY: Vintage Books, 1986), p. 45.

50 Crapanzano, p. 45.

51 Crapanzano, p. 44.

52 Crapanzano, p. 45.

53 'c'est toujours la même historie, les jours se succédant sans qu'interfèrent les saisons […] [s]auf quand passe un cyclone, ou qu'on volcan mugit comme une vache au pis gonflé […] [s]auf quand la terre tremble'. Martin, p. 54.

54   Le premier matin a un air de commencement de monde. Le 13 janvier, le terrain de tennis se vide dès l'aube. Hagards, certains sont accrochés dans un tissu d'illusions. Paralysés par la lumière du jour, ils traînent près de la piscine désemparés. D'autres déploient une énergie folle pour ne pas avoir l'air vaincu. Ils se déplacent et refusent d'admettre le fait que la terre a tremblé.

Saint-Éloi, p. 49

55 'Le soleil … Imperturbable, il se lève déjà sur la ville. Mercredi 13 janvier. 8h15. […] C'est un matin clair, gorgé de promesses. Chacun cherche une raison de ne pas s'abandonner au désespoir.' Saint-Éloi, p. 57.

56 'Certains nient l'ampleur de séisme, accomplissant avec le même entrain les gestes quotidiens.' Saint-Éloi, p. 64.

57 'Le jour d'après est pire. On a appris la veille que plus rien n'a de sens. Même les phrases qui font croire en la solidité de la terre.' Saint-Éloi, p. 120.

58 'La nuit du 13 janvier ressemble à celle de la veille, avec la familiarité du désespoir. Les cauchemars. Les pleurs. Chacune des répliques informes que désormais rien ne sera comme avant.' 'The night of the 13th of January resembles the night before with its familiarity of despair. Nightmares. Cries. Each aftershock states that nothing will be like before.' Saint-Éloi, p. 125.

59   Je me demande dans quels romans ou dans quels films j'ai déjà croisé ces ombres fuyantes. Ces villes défaufilées par la catastrophe. Je ferme les yeux et me rappelle certaines scènes de *L'Aveuglement* de José Saramago, de *Trilogie new-yorkaise* de Paul Auster, de *La Route* de Cormac McCarthy. Désastre. Désolation. Paradis perdu.

Saint-Éloi, p. 63

60 'Ces paysages auxquels j'appartiens ne sont pas les scènes d'une fiction.' Saint-Éloi, p. 63.

61   Comment pleurer un proche quand on ne sait pas vraiment s'il est décédé? Comment pleurer quand secrètement on attend le retour de celui qui est sorti le matin? On l'attend le soir du 12 janvier et les jours à venir, à chaque bruissement de pas, à chaque réplique. On risque d'attendre *ad vitam*.

Saint-Éloi, p. 62

62 The distinction between mourning and melancholia is based on Freud's now classic essay 'Mourning and Melancholia'. Nouri Gana summarises Freud's argument:

While mourning is a normal affect that is accomplished once all object-cathexes are withdrawn from the lost object and displaced onto a new object, melancholia results from an unfaltering fixation on the lost object and culminates in a regressive process of incorporating, if not devouring, the lost other – a process that might eventually enact a primary narcissism, and that Freud suspects of being of a pathological disposition. Thus, whereas in mourning the lost object is remembered so as to be consciously knitted, in accordance with the commands of reality, into the texture of the psyche, in melancholia the object is unconsciously engraved within the psyche.

> Nouri Gana, 'Remembering Forbidding Mourning: Repetition, Indifference, Melanxiety, Hamlet', *Mosaic: A Journal For the Interdisciplinary Study of Literature*, 37 (2004), 59–78 (p. 60)

63 Evelyne Trouillot, 'Eternity Lasted Less than Sixty Seconds … ', in *Haiti Rising: Haitian History, Culture and the Earthquake of 2010*, ed. by Martin Munro (Liverpool: Liverpool University Press, 2010), pp. 55–9 (p. 55).

64 Laura Wagner, 'Salvaging', in *Haiti Rising: Haitian History, Culture and the Earthquake of 2010*, ed. by Martin Munro (Liverpool: Liverpool University Press, 2010), pp. 15–24 (p. 23).

Part III
Disaster space

6 Navigating through the rubble

The immediate events associated with the January earthquake extended well beyond the confined moment and also the location of their occurrence. They intervened and were embedded in a complex context of environmental inter-actions that were shaped in turn by socio-economic and political processes. The January 2010 earthquake was a highly specific disaster characterised by its post-poned future impact yet bearing some resemblance to past geological ruptures.[1] Narrative treatments of the tremors, such as Dany Laferrière's *Tout bouge autour de moi* (2011, henceforth *Tout bouge* and translated as *The World is Moving Around Me*), a series of 128 impressionistic sketches revised in the text's sub-sequent editions, and Sandra Marquez Stathis's memoir and travelogue, *Rubble: The Search for a Haitian Boy* (2012, henceforth *Rubble*),[2] understandably focus on the moment the earthquake struck and its immediate impact. At the same time, however, they evoke the disaster's various pasts and recursive histories and contemplate their significance for future times and for potential visions of indi-vidual and communal remaking and the city and the country's reconstruction.[3] They share in their respective, and highly contrasting, attempts to forge a nar-rative form and imagery that would approximate the experience of the 'before' and 'after' of the earthquake space. In the process, first, they demonstrate the differentiated experience of rubble and ruins, as absence and loss or presence and survival, exposing asynchronous temporalities of remaking, reconstruction and rebuilding. Second, they explore the central issues of post-disaster literary endeavours, namely: how to best represent and share the experience of the highly particular yet context-specific event and its unparalleled, life-redefining material impact without reproducing and rehearsing the very images and discourse on Haiti's uniqueness, and its unique ruination, seen only in extreme terms.

Both loosely autobiographical accounts, *Tout bouge* and *Rubble* attempt to confront the omnipresent material devastation and creatively consider the histo-ries of different urban and rural sites, contemplate the unfolding histories of local island landscapes – 'the slavery [visible] in the vegetation'[4] – that is the colonial pasts and postcolonial present of island ecologies, engaging with the historical, political and ecological processes that determined the scale of the January 2010 disaster. In their response to the rubble, ruins and ruination, the two texts expose the contrasting meanings of pre- and post-disaster space, the here and now, while pointing to the differentiated logics and temporalities of remaking – an individual/collective work of affect and care – and rebuilding and reconstruction, often still charted as a teleological process and an accelerated time. Whereas the former, following Katrina Schlunke's distinction, calls for the possibilities of 're-envisaging home as affective assemblage',[5] the latter is marked by 'ruination of teleology'[6] and 'progressivist thinking' which assumes that '[o]nly rebuilding signals "progress"'.[7] Within this differentiation, rubble 'can be seen and removed as quickly as possible'[8] but a ruin cannot. At the same time 'ruin', along with its connotation of an 'object's pastness'[9] and its value as heritage, is often positioned in opposition to the value-less rubble, in an 'attempt to conjure away the void of rubble and the resulting vertigo that it generates'.[10] Rubble, however, 'is not simply a figure of negativity';[11] it 'exerts positive pressure on human practice and is constitutive of the spatiality of living places'.[12] In short, rubble and ruins are not 'nothing'. They are everything for affected individuals, families, and communities have and can hold on to as they continue to look through the debris in search of minute fragments – reminders of a life before the minute the earthquake struck – while trying to form a new 'affective assemblage'[13] of care for all that and all those who remain. In their respective attempts to convey the scale of devastation and the labour of these daily efforts, *Rubble* and *Tout bouge* expose the many meanings of ruin, rubble and debris and the asynchronous temporalities and split locations of reconstruction, remaking and rebuilding efforts, whether imagined or already materialised.

Otherwise formally, aesthetically and thematically distinct, both narratives point to the limitation, or even impossibility, of mimetic representation of the disaster, which, as a personal and communal experience, can only be pointed to, indexed, but never fully grasped. Trying, nonetheless, to relate the January disaster as a process to its wider chronologies and structural, systemic aspects, the two texts intertwine their respective narrators' past memories of Haitian landscapes with reflections on the contemporary state of devastation as well as the challenges to the varied remaking and reconstruction efforts. Similarly, *Rubble* and *Tout bouge*, albeit motivated by radically different personal relations to Haiti, attempt to formulate a committed response, in the face of undeniable devastation and suffering. Whereas Marquez Stathis's account as a *pledge* of commitment is vocal about the narrator's emotional and material investment in the titular Haitian boy's post-disaster future that she hopes to facilitate, imagined through his social ascent and finding a way out of the rubble, Laferrière's collection expresses its commitment, *la prise de position* (the taking up of a stance or one's position), through the very form it takes. Caught in the January disaster during one of his visits back to Haiti, the narrator of *Tout bouge* positions himself in and among the ruins as an immersed, and at times overwhelmed, observer and listener, attempting to capture the city's polyphony.

In the process, both narratives draw on the dominant geographical discourses, which oscillate between the pastoral and the post-apocalyptic, that have been repeatedly mobilised in conceptualisations of rubble, ruin, retreat and pre- and post-earthquake Haiti and its urban and rural environments. In triadic terms, these connect the past to the pastoral retreat, equate the present with devastation and omnipresent and menacing rubble, and envisage the future as being marked by a sense of enduring ruination and inertia that, if at all, can only be countered by the telos of rebuilding a bucolic refuge. Yet, despite their potential simplifying effects, pastoral and post-apocalyptic, and the contrasting connotations of 'idyll/retreat' and 'ruin' they carry, allow, at the same time, for a reading of Haitian geography that explores the material conditions of the varying sites as well as the processes that shaped them. In this sense, both 'retreat' and 'ruin' function as a verb and a noun, referring to a state of a place in a given time as well as the processes that made it so. In their consideration of the longer histories and futures of Haiti's urban and rural sites as well as the wider ecologies within which they are embedded, Laferrière and Marquez Stathis attempt to respond to the visible destruction and conjure a vision for Haiti's possible reconstruction, contemplating, in the process the past, present and long- term future of the island's geography, seen in both temporal, discursive and recursive terms. This last view of 'natural' history of disaster as recursive emphasises, after Ann Laura Stoler, the ways in which 'these histories are marked less by abrupt rupture or by continuity' and are rather 'processes of partial reinscriptions, modified displacements, and amplified recuperations'.[14] Such a move away from binary notions of rupture or continuity allows the readers to account for the varied forms and non-linear processes that ruin and retreat, in their dual mode as nouns and verbs, call forth and re-inscribe.

Moreover, the two texts reposition the 2010 disaster and its visible aftermath as both an individuated site of subjective experience on which subjectivity is founded and a multi-layered, collectively experienced ecology, a densely constructed network through which the interconnected social, historical and environmental coordinates of the disaster can be creatively charted and explored. In their divergent approach, both *Rubble* and *Tout bouge* attempt to negotiate the tension between the need to testify to and report from the site of devastation and the impossibility of mimetic representation of disaster; they can only describe their here and now, indexing the wider experience of the catastrophe. Finally, the two texts under scrutiny both show how literary narratives can contribute to our understandings of the 2010 disaster. First, they allow the readers to enter and understand the logic in which oft contrasting notions of reconstruction, rebuilding and remaking operate. Second, they enable a critical examination of the types of narratives on Haiti's *natural histories* that underlie these different visions – histories that are sometimes occluded, concealed as much as they are exposed – their material manifestation in a non-harmonious, uneven and convalescent world.

Quotidian ruins

Anchored in the exploration of the urban everyday and diurnal efforts to navigate the ruined city, Dany Laferrière's *Tout bouge autour de moi* attempts to find the 'right' language and form that would allow the narrator to express, reflect on and, eventually, come to terms with the scale of devastation caused by the earthquake as well as comprehend the significance of this historic rupture. More vocal about the impossibility of achieving a full grasp of events – an understanding that might be reflected in a more controlled linear narrative form – Laferrière chooses instead to emphasise the fragmentary nature of his account. A collection of short vignettes, that in themselves are unfinished reconstructions and works in progress, *Tout bouge* is explicitly concerned with the urban devastation the disaster caused, with its immediate implications for Haiti's reconstruction, and with the ways – as much limiting as enabling – in which both the damage and its aftermath have been discursively framed.

The aesthetics of Laferrière's collection suggests the impossibility of establishing, at least initially, any comparisons or clear links between the before and after of the earthquake. The 2011 La livre du poche French edition of the text, which is the primary edition for my analysis, starts with 'La minute' ('The Minute'), the life-changing minute the earthquake struck, with the entry clearly announcing a violent and uncanny manifestation of a historical moment: 'Suddenly we saw a cloud of dust rising into the afternoon sky.'[15] The revised and extended Mémoire d'encrier edition,[16] for its part, opens with the more affirmative 'Déjà la vie' ('Life Returns') entry that seems to emphasise the community's resolve, despite all, to carry on with their daily lives, with the reverberating sounds of laughter, not tremors, filling the air: 'Life seems to have gotten back to normal after decades of trouble.'[17] Following on from this hopeful image, the collection returns to the moment the earthquake struck, as captured in 'La minute'. While the opening entries indicate a shift of emphasis, the closing chapters of the two editions further demonstrate the iterative character of Laferrière's narrative of the January earthquake and its impact. Both his story and the history of the disaster emerge as open-ended works in progress, subject to further revisions and rewriting. After 'La tendresse du monde' ('The Tenderness of the World'), which is the final chapter of the Livre du Poche edition, the Mémoire d'encrier edition includes an epilogue section, composed of five, meta-thematic entries that describe the physical and creative itinerary of the account, as the author-narrator moves between Port-au-Prince, Montréal, Bruxelles and Paris. In as much as they add to the self-reflexivity of the whole collection, these additional entries also recall and echo the life-affirming opening reverberations of laughter with the music of a couple's amorous encounter providing a final note to Laferrière's narrative composition. Still, this life-giving cadence does not fully mute the uniquely haunting roar of *goudougoudou*.

It is the revised and extended version of *Tout bouge* that was the basis for the English translation and edition, with supplementary extratextual features adding a layer of mediation and self-reflexivity to the text. First, it seems significant that in the English version, the title and the supplementary components of the book suggest a generic qualification of Laferrière's collection: *Tout bouge autour de moi* becomes *The World is Moving Around Me: A Memoir of the Haiti Earthquake* (2013). This sub-title, as well as the summary information provided on the back cover, which describes the text as a 'revelatory book [and] an eyewitness account',[18] accentuate the importance of the narrator's own direct experience of the event. In so doing, they reframe the collection so as to bring the disaster and the story of one's survival, given more prominence in this framing, to a wider audience who, possibility less familiar with Laferrière's *œuvre*, might have followed the event from afar as it unfolded on TV screens. At the same time, the sub-title also seems to be a marketing ploy in response to the growing popularity of memoirs in the North American publishing market and the 'memoir boom'[19] of the last decades, poignantly dubbed 'the era of witness'.[20]

The summary further posits a surprising, if not unproblematic, link between Laferrière's account and earlier post-Holocaust testimonial narratives: 'In this way, this book is not only the chronicle of a natural disaster; it is also a personal meditation on the responsibility and power of the written word in a manner that echoes certain post-Holocaust books.'[21] By creating this parallel and suggesting a possible comparison between the two events, the cover information takes away a degree of specificity from the event, stressing instead the importance of the transferable cross-cultural insights the autobiographical narrative offers. Likewise, a foreword by Michaëlle Jean (former Governor General of Canada and Special Envoy for Haiti for the UNESCO) as well as the inclusion of a meta-thematic closing section, 'How It Came to Be', both again absent from the first French original, widen the appeal of the account by presenting it as a generic example of literary testimony. In effect, the foreword and the title of the English edition create a set interpretative frame that seems to work against and works towards an ordered bringing together of the fragmentary and jumbled elements of the original text which precisely resists inserting the earthquake within a relational narrative frame.

Although *Tout bouge* does begin with the moment the earthquake struck, in contrast to the framing of the English edition, there is no overarching theme or goal to the collection. On the contrary, it points to the futility of such literary ambitions in the context of an event that can neither be represented in its entirety – it is still ongoing – nor fitted neatly into a pre-determined narrative frame. Throughout the text, the narrator-author figure draws repeated attention to the limitations of narrative representations of the earthquake which are additionally coloured by the kinds of subjective vision and emotional investment that are the stock-in-trade of first-person autobiographical accounts. What the collection *can* offer, however, is a series of encounters with people, objects and situations that the narrator makes as he tries to navigate the new city space and to make sense of what has happened. He is not interested in his work becoming an authoritative guide: rather, his entries, which have a deeply intimate and self-contained character, neither reach out directly to the reader nor make any attempt to provide a broader rationale for the account. Instead, in all its complexity, *Tout bouge*, in Fabienne Pascaud's gloss, offers 'a surprising memento'[22] – at once a summary, a reminder and a souvenir of the 'past event or condition, of an absent person, or of something that once existed'.[23]

Despite the differing emphases and generic qualities suggested by the formal framing of the two language editions, the text of the account itself is conspicuously fragmentary in design. The mosaic-like form of the work, as an assemblage, acts as an overarching and recurring meta-thematic comment on the nature of the event, 'which burst upon its victims with no prior warning and completely altered their lives in less than a minute',[24] and the necessarily limited view it offers, thereby reinforcing the narrator's general criticism of totalising representations of Haiti over time: 'I get the impression that everyone is using the same bank of images. Within two hours, I saw the same closed expression on a little [girl's face] standing in a crowd a dozen times.'[25] The comment is a direct challenge to the stultifying character of mainstream media portrayals of Haiti, and it is followed by an equally bracing reminder that some of the 'expected' scenes of chaos did not occur: 'In the end, there never were those chaotic scenes that some journalists (but not all) no doubt wanted to see.'[26]

Here as elsewhere, the narrator questions reductive readings of the disaster and refuses to subscribe to the dominant media image of Haiti as a site of enduring violence in which the most recent episode of material devastation comes to embody the general, 'permanent' state of Haiti's collapse. What the narrator remarks on briefly here, the American journalist Jonathan M. Katz, who also survived the earthquake, analyses more extensively in his book-length study of the inadequacies of foreign aid and the ideological assumptions behind it. As Katz argues, if in somewhat simplified, triadic terms, negative images of Haiti had a similarly adverse impact on the way in which post-disaster relief and the recovery effort were directed: 'Having sought above all to prevent riots, ensure stability, and prevent disease, the responders helped spark the first, undermine the second, and by all evidence caused the third.'[27] In effect, both the fragmentary form and explicit narratorial commentary contribute to *Tout bouge*'s self-reflexivity and work against totalising depictions of Haiti as a ruin that preclude a critical engagement with the country's past, present and future.

Laferrière's criticism of stock disaster imagery underlines how the predictable portrayal of Haiti's 'extreme violence' and extreme ruination, paired with discourses on the 'natural and innate resilience' of its population, is tinged with exoticism, presenting Haiti as essentially 'unknowable', an irrational site, and, in more negative versions, the Haitian people as a 'disorderly mob'. Such vocabulary, also underpinning some of Marquez Stathis's depictions of urban debris as discussed later, thinly disguises racial essentialism and reaffirms the construction of Haiti as a peripheral and *unmodern* site of exceptionality and singularity.[28] Challenging uses of 'resilience' and 'ruination' as antithetical and trans-temporal categories, Laferrière insists on the *specificity* of the 2010 earthquake, which in one sense could have happened anywhere, but whose multiple informing contexts clearly differentiate it from other, similar but non-identical global events:

> And it wasn't because of a coup or one of those bloody stories mixing voodoo and cannibalism – it was because of an earthquake. An event over which no one has any control. For once, our misfortune wasn't exotic. What happened to us could have happened anywhere.[29]

In this extract from 'L'instant pivotal' ('The Pivotal Moment') the narrator explicitly contests the dominant rhetorical frames, e.g. those attributing the earthquake to 'Voodoo' or 'Haiti's culture', that were used to account for the disaster. Such bogus 'explanations' are of course anything but new. For the narrator, they are rehearsals of earlier responses to an equally important moment, the 1791–1804 Revolution, and the establishment of the world's first black republic – an inconceivable and unacceptable event for the world's major colonial power – in response to which the world would turn its back on Haiti, a punishment that (for Laferrière) would last 200 years. Drawing on these images of Haiti as an 'extreme space', where the earlier revolutionary impulse is recast as a source of the country's current predicament, post-earthquake claims of the country's unique suffering and corresponding 'natural resilience' feature here almost as a remedy and justification for the 'natural', rather than acquired, compound vulnerability of postcolonial island populations such as Haiti's.[30]

Where socio-scientific analyses are necessarily focused on environmental interconnectedness and the compound nature of vulnerability, privileging a removed and structurally-focused, bird's-eye view perspective, Laferrière's collection provides an insight into the quotidian and personal experience of these socio-ecological phenomena and their lived, micro-scale manifestations. Through a non-hierarchical bringing together of multiple perspectives, with the narratorial 'I' being only one of many voices involved, *Tout bouge* invites a reflection on compound socio-ecological vulnerability, articulated less as a set of identifiable intertwined factors or a structural condition and more as a day by day struggle against the lack of or differentiated access to essential and life-sustaining resources. In 'Le bois' ('Wood'), for example, the narrator talks about deforestation via the difficulty of acquiring building materials,[31] whereas in 'L'électricité' ('Electricity') he portrays his family's joy, which for him seems almost irrational, for the few hours a day when the electricity is on. Finally, in 'Le golf' ('Golf'), the narrator transitions from general observations on a displaced population to a commentary about the politics of unequal density in Port-au-Prince. The city's golf course, ordinarily a status symbol for the capital's elite, was transformed into a makeshift camp for earthquake survivors, and this specific site functions as an ironic reminder of the jarring socio-economic disparities that were mapped onto urban space long before the disaster struck. In his characteristically unemotional manner, the narrator juxtaposes the image of the overcrowded camp with the vast space required to play the game – one that seems inappropriate for such a populous city:

> The game is hard to comprehend in a city so overpopulated. It takes up too much space for too few people [...]. A tiny white ball for such a vast surface – it seems like another provocation.[32]

In the closing sentence, the narrator highlights this tension by contrasting the minuscule white golf ball with the vast green terrain, its fertile soil squandered for leisure despite the lack of arable land: 'The land is good, but there's not a single fruit tree in sight.'[33] The use of space for luxury leisure pursuits is starkly opposed to the real labour required to farm land; excluded from this site of privilege and recreation, Port-au-Prince's inhabitants can only reclaim the space temporarily, with each day marked by a threat of eviction or displacement to another camp.

Yet even in this direct commentary that established key challenges to the city's reconstruction – namely unequal access and distribution of land, poor infrastructure and need-driven ecological exploitation – there is no claim to authority; rather, the narrative provides fleeting, everyday glimpses of systemic fragility and frustrated individual attempts to mitigate it. Capturing the quotidian and micro-scale translations of socio-ecological vulnerability and inequality, the text points to the ways in which histories of landscape use and exclusions translate into and shape everyday efforts to rebuild a life among the ruins. There is no guiding narrative voice that authoritatively interprets post-earthquake realities or claims to provide a solution to complex issues. Instead, by loosely assembling his own observations, combining these with strangers' passing comments and family conversations, the narrator offers a mixed perspective on the earthquake, the factors determining the scale of its aftermath, and the plethora of responses it provoked.

Labyrinthine rubble

In *Rubble: The Search for a Haitian Boy* by Sandra Marquez Stathis, the contemplation of Haiti's pre-earthquake past, present devastation and future forms of reconstruction takes the form of a personal account of the narrator's different journeys through Haiti's urban and rural landscapes. Marquez Stathis's narrative interweaves past and present reflections on Haiti's landscapes, a move that takes the reader back to the early 1990s and posits the earthquake in relation to the narrator's earlier life-events. The narrative is imbued with bucolic imagery that accompanies the narrator's first, fondly remembered trip to Haiti in the 1990s, and these nostalgic images then come to define the text's oppositional representation of Haiti's landscapes both before and after the earthquake struck. On the face of it, *Rubble*, it seems, is a conventionally structured, linear travelogue written by a former UN human rights observer who was part of the newly set-up Organization of American States–United Nations Mission and a foreign correspondent for Reuters.

As its title already foreshadows, the aim of the book is two-fold: on the one hand, it attempts to provide a detailed account of the earthquake's impact; while on the other, it traces Marquez Stathis's personal quest to find Junior, the titular Haitian boy. At different points in the text, the narrator explains how her first assignment in Haiti back in the 1990s was an unforgettable adventure: one she had decided to embark upon, without much forethought, accompanied by fellow human rights and development workers. Marquez Stathis then returns over a decade after her last trip, spurred by the news of the January 2010 earthquake, in the hope of finding and helping Junior – a young boy she had previously befriended and cared for during her last stay. By contributing to his well-being and safety, she aspires to participate directly in the post-disaster reconstruction efforts. This most recent trip is also an occasion to return to the urban and rural sites that had previously shaped her perception of Haiti, and, consequently, to revisit the memories these places elicit as well as the knowledge they impart, forcing her to confront the pre-history of the disaster and the contexts underlying it. Joined together, the two narratives of her past and recent trips to Haiti create a composite travel-memoir that gives shape to the narrator's observations on Haiti's changing landscapes, their past, present and future, and the accompanying journey of self-exploration through the maze of rubble or the lush tropical garden.

The two-fold aims of the text are further reaffirmed in the book's extratextual features – a map of Haiti, a photograph of Marquez Stathis and Junior, the latter's drawings, the synopsis and reviews on the back cover – as well as *Rubble*'s bipartite division that signal its goal to provide an accurate and intimate testimony as well as to reconstruct and document the narrator's experience of the recent events. For example, in both the reviewers' comments and the book's summary, the emotional honesty and intensity of the account are coupled with the credibility of the text – frankness is translated into accuracy and authenticates the narrative. In this context, the author-narrator figure occupies multiple positions as 'a mother, journalist, and former human rights observer', 'chronicl[ing] her soul-etching and life-altering years living and working in Haiti'.[34] In its two-fold function, *Rubble* is seen as at once a 'chronicle', a register of events 'without philosophic treatment, or any attempt at literary style',[35] and an 'eyes-wide-open portrait of Haiti'[36] – a creative work coloured by one person's necessarily limited perception.

This importance of direct sensory experience, the 'eyes-wide-open-portrait of Haiti', is further emphasised through the bipartite division of the book into Part I – 'Je/Eyes' – and Part II – 'Ké/Heart'. Each is preceded by an epigraph: Mary Catherine Bateson's *Composing a Life* (1989)[37] opens Part I and Jorge Luis Borges's 'Two English Poems' (1934),[38] Part II. These foreground the narrator's aspirations, mark the two stages of Marquez Stathis's account, reiterate the task she has set herself and point to the primary method – emotional honesty – she will employ in order to fashion and give meaning to her text. The 'authenticity' of the narrator's personal account of her journeys in Haiti and her friendship with Junior thus becomes a guarantor of the text's 'accuracy', an equivalence stressed again in the latter part of the synopsis: 'Marquez Stathis is determined to find out if he [Junior] has survived the quake and to express her gratitude for his enduring friendship. In so doing, she learns that Junior's story *is* Haiti's story' [italics added].[39] Within this formulation of the narrative's aesthetics, 'chronicle' and 'portrait' are perceived as complementary threads that, despite their different functions, can nevertheless be seamlessly interwoven into the fabric of the text.

Narrative focalisation throughout the account is thus used to bring together and stress the complementarity of the two elements of Marquez Stathis's travelogue. Yet although the single narrative voice provides such unifying frame and successfully contributes to the sense of intimacy between the narrator and the reader, it effectively limits the perspective on Haiti that the text offers. With no countervailing viewpoint or voice being incorporated into the account, the reader is only exposed to and immersed in the narrator's single interpretative frame. Privileging this focalised and firmly situated perspective, Marquez Stathis uses the honest, if less self-reflexive, appraisal of her own experience of Haiti as the basis for a generalising analysis of the country's predicament in which this pairing of emotional honesty and authenticity – central features of memoir – paradoxically limit rather than increase the reliability of the contextual account. In sum, together, *Rubble*'s extratextual features and its form suggest a sense of complementarity between the two main narrative threads and formal conventions, implying, in effect, that emotional honesty in its sincerity and intensity translates directly into a form of detailed yet comprehensive narrative impartiality – a tenuous pairing of a deeply personal view and a panoptic perspective on the disaster that the text hopes to create.

Moreover, the repeatedly reaffirmed truthfulness of the account is directly linked to the text's ambition to counteract stereotypical perceptions of the country. Aware to some degree at least of the ways in which Haiti and its people have been exoticised, the narrator does not want her memoir to do the same. Rather, she sees her bipartite narrative, precisely, as an intervention against such objectifying readings:

> [...] I have tried to be judicious, never revealing more than needed to be told and never holding back in the service of emotional honesty.
>
> I have tried to take that same approach in writing about Haiti and her people, looking beyond the accepted narrative of violence, chaos, misery, and tragedy – and instead looking deeply into Haiti's mirror image to source our common humanity.[40]

These consecutive affirmations, in the book's acknowledgements, provide a direct meta-thematic and authorial commentary on 300 pages that precede them with the repeated phrase, 'I have tried', preemptying the reader's potential assessment of the text. As such, the authorial remark also points to the core challenge of the narrative, which she does not fully overcome at any point in the account, of joining her emotional and analytical inquiry into the contextual and deeply personal meaning of the disaster and its wider resonance.

As its title already suggests, the overpowering presence of debris characterises the narrative portrayal of post-earthquake urban destruction in *Rubble*. Haiti's capital, following the earthquake, is a benighted place of permanent ruin and primeval darkness, in stark contrast to the lush bucolic greenery of the Haitian countryside. The emotional charge of the narrator's observations reflects her shock at the scale of the devastation of approximately 80 per cent of the capital. From the prologue onwards, rubble accumulates relentlessly to present an image of Port-au-Prince as a post-apocalyptic space. Here are her first impressions of the country from the opening paragraphs of the book's prologue:

It goes on and on for as far as the eye can see. Rubble, a mass grave of final breaths and buried dreams. Rubble, a newly formed tundra of boulders, concrete slab, mangled metal rods, detritus of daily life, and cinder blocks returned to their natural state of cement ash. Rubble, a shifting archipelago forever altering the topography of land and memory.[41]

Her later observations of the city space, unlike those of the recovered countryside, reiterate these images of omnipresent decay:

In its stead are heaping piles of trash, a monument to the presidential stalemate paralyzing the country right now. The sky has grown overcast and the sight of all the trash weighs on me. I'm not sure if it's just a matter of my eye needing to adjust to Haiti or if the country has slipped ever deeper into the muck.[42]

In the first quotation, the debris reaches 'as far as the eye can see' and has an enduring, timeless character to it. Debris is at once a category of space and a property of experience. The scene in which the narrator is immersed resembles a labyrinth of concrete, metal and cement ash that she, a latter-day explorer-figure, attempts with great difficulty to navigate. In contradistinction to earlier representations of Haiti's rich and fertile flora and fauna, and to the associations of peace, simplicity and purity these evoke, the natural formations recalled here (e.g. tundra) have no such positive and generative connotations. Rather, these portrayals amplify a sense of overpowering density that exacerbates the violence and chaos of an afflicted and flagging city.

Rubble's insistence on the omnipresent and all-encompassing devastation of the capital seems, paradoxically, to reposition the earthquake as an urban, 'non-natural', event conceptualised primarily through the visible material devastation of the capital that it had caused. Destruction, accounted for in characteristically emotive hyperbolic terms, is total and permanent: Haiti is now configured as a *discarded* as well as a *devastated* space. Yet, narrative portrayals of ruination can also serve to conceal as much as to excavate the fissures that divided the city space and effectively bled Haiti's body politic. For one, the terms 'ruin', 'ruination' and 'devastation' can alternately refer to specific depictions of post-earthquake urban landscape and to general discourses of underdevelopment that characterise Haiti as a 'failed state' (Part I). And for another, their temporal signification may connote either a sense of permanence and impenetrability or point to the long-term processes that contributed to a state of ruination. 'Ruin', as Ann Laura Stoler explains in her analysis of imperial debris, can be both a 'claim about the state of a thing and a process affecting it'.[43] 'Ruin', after all, functions as a verb as well as a noun; it is 'the state or condition of a fabric or structure'[44] or a person, but also an action of inflicting 'great and irretrievable damage, loss, or disaster upon (a person or community)'.[45] 'Ruins' and 'ruination', in this wider sense, can refer to three distinct moments across time, showing how past, present and future are all shaped by a range of 'violences and degradations that may be immediate or delayed, subcutaneous or visible'.[46] This wider definition teases out and engages with the significance of narrative portrayals of ruins and ruination, both physical and symbolic, for Haiti's past, present and future.

In the second observation, post-earthquake Port-au-Prince is depicted as a place that has, after multiple blows, eventually 'collapsed into a state of prehistoric rubble'.[47] An overriding sense of blight, immobility and ongoing ruination, as a continuing and unstoppable process, permeates the passage, conjoining images of physical decay and political paralysis – lacklustre metaphors of an inert and ruined state. Meteorological imagery further contributes to this anguished picture of Haiti as a discarded and devastated place, its urban landscape desolate and empty. The scene seems timeless or, perhaps better, to exist in jumbled time, with the country being presented not only as the site of a recent disaster but also of other disasters waiting to happen, other catastrophes inevitably to come. Within this immobilising view of unending ruin, Haiti, in Greg Beckett's gloss, 'is rendered pathologically stuck in a temporality of crisis, in a time that goes nowhere'.[48] Although these highly charged passages are presumably aimed at triggering a sympathetic response on the part of the reader, they ultimately risk doing the opposite, producing a disempowering effect. For the shocked and overwhelmed narrator, the whole country has been transformed into a dystopian wasteland, a paradigmatic 'Disasterland'[49] with the neologism becoming a synonym for Haiti itself.[50] Within this binary characterisation, which clearly aims to illustrate the traumatising force and the emotional impact of material devastation on the narrator, Haiti is reconfigured as a contradictory space, 'a brainteaser',[51] a place where 'expectation and reality so often clash'[52] and where 'reality is a constant puzzle, existing on multiple layers'.[53] Furthermore, the hyperbolic diction and imagery of Marquez Stathis's observations, in language uncannily, and possibly unknowingly, echoing Joseph Conrad's *Heart of Darkness* (1899), risks portraying Haiti as a toxic residue; the leftover product of a chronic process of decomposition and ruination, it is a godforsaken place that has 'slipped even deeper into the muck'.[54] The use of the word 'muck', which denotes '[m]ud, dirt, filth; rubbish, refuse',[55] along with the familiar light versus dark binary, creates a disturbing image of Haiti as a helpless and devastated place, immobilised in, at once, seemingly natural yet self-imposed vulnerability. No change, whether political, social or environmental, seems possible in this failed state, this political-cum-environmental 'disaster zone'.[56]

Presented in *Rubble* as being some way short of a modern nation, Haiti is trapped in darkness also in more political terms and needs to be guided towards the light, which is somewhat unsurprisingly embodied in the text by the narrator's own home country, on one occasion referred to as 'a beacon of hope to so many around the world'.[57] Otherwise more openly critical and reflexive about Haiti's role and significance in regional and international politics which contributed to the ruin of the Haitian state, *Rubble*'s hyperbolic depictions of devastation undermine any such more nuanced narratorial reflection. Instead, they set up the United States, uncritically so, as the 'truly' modern, democratic society and the epitome of what democracy should look like. In an extension of the binary imagery, the text sets up an antithetical and hierarchical relationship between Haiti and its powerful neighbour, one that equally draws on the American metaphor of the nation as 'a city upon a hill'. The image first used by Jesus in the Sermon on the Mount[58] and then famously employed by John Winthrop in his *A Modell of Christian Charity* (1630),[59] has since been used in political discourse to signify the idea of American exceptionalism. In effect, a simple translation of the trope, the narrative's hyperbolic imagery effectively conceals Haiti's proximity as well as its centrality in terms of both its geographical positioning and the role it played in the histories of the Caribbean and the Americas, with the Haitian Revolution (1791–1804) functioning as a founding *modern* event.[60] The first geographical aspect challenges the notion of Haiti as a periphery. The Caribbean Sea functions in symbolic and physical terms as both a separator and a link between the insular states and the U.S., which has now become a dream destination for many Haitians. The sea is a site of 'an unceasing traffic [...] that joins and isolates the first Republics in the New World'[61] and this continuous movement, epitomised in the contemporary context, in the image of 'boat people',[62] can be traced across time and space. Meanwhile the second aspect, the Revolution, a radical realisation of the ideals of equality, fraternity and liberty, was an unprecedented challenge to French rule, the global colonial system and the resulting economic supremacy of the colonial powers. A singular political and symbolic undertaking that resulted in the establishment of the world's first free black republic, the event was an act of reclamation enmeshed in the global histories of imperialism and the developing colonial market. Indeed, from its revolutionary inception Haiti has been at the *heart*, not of darkness but of global socio-political transformations.

In sum, on the formal level, such hyperbolic diction, imagery and binary formulations add to and further complicate the mixed character of the text and the tone of the narrative voice which oscillate, on the one hand, between the shocked and overwhelmed narrator providing an intimate and focalised emotional account and, on the other, the enlightened narrator, a conflicted yet still hopeful former political observer, guiding the reader through Haiti's bleak reality. Consequently, while revealing the scale of the narrator's confusion and the sense of a personal and collective rift caused by the event, the text's imagery and diction translate into an oppositional presentation of Haiti's past and present as well as a conflicting vision for its future.

A ruined state

Offering otherwise a set of very different personal reflections on the everyday experience of devastation – as the immediate post-earthquake present and as a conjuncture and manifestation of long-term and compound socio-ecological processes – both *Rubble* and *Tout bouge* embed in their accounts a contemplation on the material and symbolic ruination of the state. Among the many sites of destruction encountered, Hotel Montana, Hotel Christopher and the National Palace, symbolic of the tension between foreign presence in Haiti and the country's sovereignty, engender particularly powerful responses in the two texts. In their respective engagements with these same urban locations, both narratives reveal the multiple, often discordant, personal and collective meanings they carry and the contrasting visions of the country's reconstruction they might inspire.

In Marquez Stathis's account, the now-ruined National Palace is depicted through the use of anthropomorphic metaphors, hyperbole and pathos that evoke images of decline and waste, and hope to produce feelings of grief and confusion: 'I can still feel the shock I felt when I saw that first crushed image of the destroyed neoclassical beauty-like seeing a bride turned into a corpse.'[63] The arresting metaphor of the Palace as a dead bride is an extension of the text's portrayal of the country as a vulnerable female friend who must be saved:

> This lovely beauty, her face cracked, her spine crushed. She can no longer stand. And yet, she is regal, dignified, and grand. Her scars perfectly mirror the pain and struggle of her noble people.[64]

This accumulation of corporeal metaphors and staccato sentences creates a double-edged portrayal of Haiti as a suffering yet dignified nation as an anguished beauty burdened by the weight of her misery. At the same time, these highly emotive descriptions seem to imply, in a move away from sympathy to more hierarchical evocations of pity, that it is precisely the country's hardship that is the source of its nobility. Both Haiti and its peoples are 'cracked' and 'crushed', yet in their vulnerability they acquire an ennobling dimension. Emotion prevails across the narrative's descriptions of this ruination; history and politics are drained of specificity and significance. At the same time, the choice of figurative language here and in other depictions of the National Palace, while justifiable in itself as a way of translating one's emotional overwhelm, sets up a tension between the alleged aims of the text and what it actually achieves; and one result is that *Rubble* ends up reinforcing the same discursive frames it had set out to deconstruct.

If the ravaged National Palace functions as a symbol of Haiti's ruin and the consecutive corrupt and, in the narrator's view, almost equally inefficient regimes[65] that led to the country's collapse, then Hotel Montana and Hotel Christopher illustrate the presence of the international community in Haiti. Dan Woolley (Part IV) is among those trapped in the ruins of Hotel Montana. The two destroyed hotels, however, are not accorded the same significance as other sites of urban debris. Rather, their ruins, unlike those elsewhere in the city, are sanctified by the heroic deaths of the foreign workers who resided there – an acquired dignity reflected in the narrator's observations. Hotel Christopher, former UN headquarters, is compared to the Acropolis,[66] and Hotel Montana, famous for its international clientele, is characterised as a sacrificial site:

> I still can't imagine how such a pillar of Haiti's development community, with so many of its citizen soldiers trapped inside Trojan Horse-style, could

crumble, collapse, and pancake into an entirely different state of matter with such devastation, speed, and efficiency.[67]

The passage employs classical references, e.g. the inaccurately used image of the Trojan horse,[68] as well as biblical metaphor in order to set the sites apart from other, equally chaotic scenes of urban debris – a distance that paradoxically reflects the actual detachment of the international community from the local population.

No longer an ordinary lodging, for the narrator, Hotel Montana has 'become a true cathedral, a gathering spot and final resting place that continues to bring people together'.[69] It is a battlefield and a site of heroic sacrifice of the 'citizen soldiers' who were trapped inside, were 'a part of a global chain', and whose deaths were '[a] sacrifice [which] mattered' and which will be built upon'[70] – a clear reference to the biblical story of Abraham and his son Isaac in the Book of Genesis.[71] It remains unclear, however, in what ways the accidental death of many might be seen as being 'for the attainment of some higher advantage or dearer object'.[72] In this emotionally charged depiction of the hotel's collapse, the text proffers vague biblical allusions with little consideration of the dubious implications these might carry. Consequently, it imparts a heroic dimension to the death of foreign workers which gives the reconstruction process a similarly quasi-sacred character reflected in the language of mythological renewal and rebirth.

In an extension of the metaphor, the hotel, risen Phoenix-like, becomes a sacrificial site yet one which, for the narrator, can still be a haven for 'Haiti's humanitarian soldiers'. They can 'continue to report for duty – and enjoy a refreshing dip in the clarifying water'.[73] Its collapse, however, like that of hundreds of other buildings in Port-au-Prince, was due to the sheer force of the seismic shock allied to inadequate building material and lack of regulation. Its reconstruction was largely profit-driven and part of a larger push for the growth of the tourism industry, and 'building back better',[74] in the aftermath of the earthquake. The whole city was levelled by the disaster; and in this respect, the Montana's and the Christopher's destruction, despite their obvious emotional significance for the narrator, were in no way different to the collapse of other civic and residential buildings.[75] Within the wider context of Marquez Stathis's narrative the spatial referents used to depict the hotels and the presidential palace function metonymically as a collective reminder of Haiti's collapse with little, if any, possibility of rebuilding the city and the civic space – a sentiment also discernible in her emphasis on the recovery of Haiti's bucolic landscapes. In effect, although the use of biblical and classical imagery, pathos and highly emotive language is meant to lend a certain dignity to these particular sites and those who died there, it effectively risks taking such dignity away from many other, ordinary victims of the disaster.

Aware of the difficulty of this task to express the unprecedented scale of loss and destruction without turning to hyperbole or dramatic imagery that risks bordering on the sensational, *Tout bouge* resists fatalistic or overly personal focus in describing the country's political landmarks. The narratorial 'I', in contrast to Jeremy D. Popkin's assertion that '[of] male survivor-authors all insistently keep themselves at the center of their narratives',[76] is noticeably absent from those entries that contemplate the multiple associations of the National Palace and Hotel Montana, among others, and the collective impact of the buildings' collapse:

> The fact that dictators have squatted here, more often than not over the past two hundred years, doesn't make that piece of furniture [presidential armchair-the National Palace] any less desirable. People have never mis-taken the building for its occupier. One day they hope to restore its splendor. A wave of feeling submerged the city, the whole country even, when we learned the palace had fallen with the first tremor.[77]

Here, despite its obvious connection with previous dictatorial regimes, the National Palace remains a collective symbol of national identity, and it is Hai-tians themselves who will 'restore its splendour'. The narrator acknowledges Haiti's difficult political history, but does not disallow the possibility of a fully functioning future state, however long this might take. Rather than simply rehearsing the rhetoric of victimhood and passivity, he seeks a different narrative of the nation, turning to contemplate the revolutionary potential this disappear-ance of material and symbolic frames offers and the difficulty of rising up to it:

> Nothing is holding us back. No more prisons, no more cathedral, no more government, no more school – it's the perfect opportunity to try something new. An opportunity that won't knock twice. The revolution is at hand, and here I am, sitting under a tree.[78]

State structures have clearly dissolved and the narrator – here as elsewhere – does not fight shy of saying this.[79] Yet the use of anaphora, which also quickens the pace of the passage leading up to a moment of climax, emphasises the para-doxical opportunities that this state of dissolution allows. There is no state, no church, no school, no government – an absence that almost has a revolutionary potential to build the country's future from the ground up, on different terms. At the same time, the quickened, anticipatory pace of the passage, its expectant and hopeful tone, is suddenly reversed and halted as the entry draws to a close: stunned by the scale of the destruction, the narrator, in the French original, 'remains sited in his corner', under a tree. Gone is the revolutionary promise and the haste for change. Instead, in a seemingly emotionally detached reversal, these are replaced by a sense of dejection and disillusionment which hint at, albeit in starkly different terms than those of Marquez Stathis's account, the slow, difficult and potentially impossible process of rebuilding self and state.

The momentary celebration of this revolutionary potential, along with a partial acknowledgement of one's own entanglement in and responsibility for its realisation, contrasts with an extended critique offered by the text of other

factors, including the unsolicited presence of foreign NGOs, which have contributed to the gradual erosion of the state and Haitians' capacity to function
independently. These, in effect, greatly decreased the state's ability to target the
population's compound vulnerability that transforms environmental hazards into
risks and disasters. Collectively disempowering, the prolonged and largely unco-
ordinated presence of parallel aid agencies and overlapping NGOs has had a
negative structural effect, with a post-crisis 'state of exception',[80] which was
used to justify further humanitarian interference, becoming an everyday reality.
In his analysis of foreign involvement in Haiti and its crippling effects on the
state, Patrick Bellegarde-Smith turns to the image of a wounded body, demonstrating that '[t]his conscious and subconscious "bleeding" of *res publica*, the
public "thing"', in order to benefit primarily the American private sector, renders
the Haitian government rather ineffectual and always at the mercy of its foreign
benefactors.[81] The blood metaphor has two intertwined uses here. First, it evokes
the collective idea of 'the body politic': 'the people of a nation, state, or society
considered collectively as an organized group of citizens'[82] and second, it
expresses a poignant critique of the economic and hegemonic underpinnings of
international intervention in Haiti, whose entire civic body is effectively 'bled'.
This is not a metaphor meant to arouse sympathy or pity, as in Marquez Stathis's
corporeal imagery, but functions as an image exposing the structural violence
inflicted on the country. For Bellegarde-Smith and others, international involvement in Haiti's internal politics has hindered rather than supported its stability
and has curtailed individual and collective political agency. This is not to say
(and Bellegarde-Smith does not say it) that all international NGOs contribute
equally to this weakening of the state. But aid workers' well-intentioned commitment and motivation still 'cannot alter the structural non-accountability and
power iniquity between their employer, the government of their host country,
and the people with whom they work'.[83]

The narrator's reading of the destruction of Hotel Montana, the subject of an
entire entry in Laferrière's collection, builds on and directly expands these
earlier criticisms, interwoven within the fabric of other entries in the collection.
While reflecting on the pre- and post-earthquake meanings of space, the narrator
offers a seemingly dispassionate yet pointed critique of the political and humanitarian power relationships the hotel has historically come to signify:

> I can scarcely imagine the disaster. All this effort to save the Montana,
> while right next to it people are pleading for help. [...] A man standing next
> to the car remarks, without much bitterness, that there's more here than the
> Montana. But it's the place where big contracts are negotiated and important
> political decisions made. The favorite hotel of the international stars who
> have gotten interested in human misery.[84]

Here, the passage reveals the dichotomy between the theory and practice of the
international aid effort that manifested itself after the earthquake in a much
higher, potentially life-saving for Woolley (Part IV), concentration of rescue

teams on these sites. The bitterly ironic tone of the last sentence targets the abuses of disaster tourism,[85] but also those of the media-trumpeted relief effort, which is profoundly out of touch with the very local people it is claiming to help.[86] As the exasperated authors of one post-intervention report urge: 'If you crave media attention and the world's spotlight, do disaster victims a favor and stay at home; disaster relief is hard enough for everyone involved.'[87] In refraining from overtly emotional language, yet without fully muting its criticism, *Tout bouge* moves away from a focus on personal towards *collective* histories and meanings of space – a change that also allows for the greater contextualisation of key urban sites.

At the same time, the increasingly detached tone of the collection, radically opposed to Marquez Stathis's overly emotional and highly metaphorical narrative, is not without its limits and limitations. Whereas, at first, it allows the narrator to accompany and incorporate multiple personal perspectives on the earthquake and its aftermath, it equally creates a growing sense of detachment and distance with quotidian observations turning into more generalised assessments of structural factors shaping Haiti's past and present. Echoing some of the recurring criticisms of NGOs presence in Haiti and its ruinous effects on the local population, Laferrière's assessment of these other causes of Haiti's metaphorical and material collapse, distinct from other more cautious statements throughout the collection, is clear and unapologetic: 'The problem is that, over time, Third-World populations have developed a welfare mentality. They know the workings of the international aid system – we can sense that. They have studied it carefully.'[88] The statement, which pushes to the extreme the detached tone of the collection, is in stark opposition to other more individualised and tentative observations that otherwise try to resist and problematise a binary positioning 'us' vs. 'them', not least due to the author-narrator's own complex diasporic sense of belonging. Even if read ironically, as precisely a critique of such blame-imparting, antagonistic discourses on aid and development, the entry still establishes a clear distancing between the narrator-observer and those 'Third-World populations' he is analysing. As such, this slight shift in the narratorial voice coupled with the sense of distancing from the national 'we' that the statement implies, introduces a level of tension within the collection with the narrator's earlier non-hierarchical position, along other voices, replaced by a more distanced, sharp gaze down at a distant, homogenised group.

In effect, this other dimension of the detached narratorial tone reveals surprising points of connection between Marquez Stathis's attempts to create a highly personal and insightful account, one that resists stereotypical portrayals of Haiti, and Laferrière's fragmentary and unemotional form which also unveils the narrator's difficulty to find and define his place vis-à-vis the affected community. Still, the text's aesthetics, driven by the attempt to capture the polyphonic experience of the disaster's past and daily remakings, try to counter the dramatic intensity of a hyperbole with a more reserved, yet no less pertinent, critique of clichéd images of Haiti always in the extremes. Against such binary positioning, Laferrière's literary aesthetic repeatedly turns to the non-spectacular, quotidian moments and daily life, stressing, at the same time, that the domestic and the global, past and present material conditions underlying the disaster and the lived experience of ruins and ruination are deeply intertwined. Finally, whereas Marquez Stathis's narrative pledge, primarily as her expression of commitment to Haiti and care for Junior, works with and emphasises an accelerated time of rebuilding and transforming the ruined present, with rubble – in the negative mode – viewed primarily as absence and loss, Laferrière's collection pauses among the rubble to take stock of the minute details of what is still there and to record the presence- and present-affirming, if mournful, polyphony filling the ruins.

Notes

1 Martin Munro provides a helpful historical overview of seismic activity on the island of Hispaniola in the five centuries since its colonisation by the Spanish. See Munro, *Writing on the Fault Line*, pp. 3–6.

2 Sandra Marquez Stathis, *Rubble: The Search For a Haitian Boy* (Guilford, CT: Lyons Press, 2012)

3 For an overview of historical planning policy as well as structural and governmental provision for urban planning in Haiti, including the extremely complex land-tenure system, see Harley F. Etienne 'Urban planning and the Rebuilding of Port-au-Prince', and in particular the section on 'The Colonial and Postcolonial Construction of Planning of Port-au-Prince'.

4 'So many things in these West Indian territories, I now begin/ to see, speak of slavery. There is slavery in the vegetation.' V. S. Naipaul, *The Middle Passage* (London: André Deutsch, 1962), p. 182.

5 Katrina Schlunke, 'Burnt houses and the haunted home', in *Housing and Home Unbound: Intersections in economics, environment and politics in Australia*, ed. by Nicole Cook, Aidan Davison, Louise Crabtree (New York: Routledge, 2016), pp. 218–31 (p. 230).

6 Schlunke, p. 222.

7 Schlunke, p. 222

8 Schlunke, p. 226.

9 Gastón R. Gordillo, *Rubble; The Afterlife of Destruction* (Durham and London: Duke University Press), p. 8.

10 Gordillo, p. 10.

11 Gordillo, p. 11.

12 Gordillo, p. 11.

13 Schlunke, p. 230.

14 Ann Laura Stoler, *Duress: Imperial Durabilities in Our Times* (Durham and London: Duke University Press, 2016), p. 27.

15 Dany Laferrière, *The World is Moving Around Me: A Memoir of the Haiti Earthquake*, trans. by David Homel (Vancouver: Arsenal Pulp Press, 2013), p. 15. 'Soudain, on voit s'élever dans le ciel d'après-midi un nuage de poussière.' Laferrière, *Tout bouge autour de moi* (Paris: Éditions Grassset & Fasquelle, 2011), p. 12. [Unless otherwise stated, all subsequent translations are from this edition. In order to differentiate between the original and the translated editions, I will reference the English text as Homel, followed by the page number.]

16 Dany Laferrière, *Tout bouge autour de moi* (Montréal: Mémoire d'encrier, 2011).

17 Homel, p. 13. 'La vie semblait [semble in Mémoire d'encrier edition] reprendre son cours après des décennies de turbulence. Des jeunes filles rieuses se prominent dans les rues tard le soir.' Laferrière, p. 9 [Mémoire d'encrier edition]; p. 12 [Livre du poche edition].

18 Laferrière. Back cover.

19 Leigh Gilmore, *The Limits of Autobiography: Trauma and Testimony* (Ithaca, NY. and London: Cornell University Press, 2001), p. 2.

20 Jeremy D. Popkin, 'Life in the Ruins: Personal Narratives and the Haitian Earthquake of 2010', *L'Esprit Créateur*, 56 (2016), 101–15 (p. 101).

21 Homel. Back cover.

22 '[…]. cet étonnant mémento'. Laferrière, back cover.

23 'memento, n.', in *Oxford English Dictionary Online* (Oxford University Press, 2015) www.oed.com/view/Entry/116329?redirectedFrom=memento [accessed 10 March 2015].

24 Popkin, p. 103.

25 Laferrière, p. 88. Homel, p. 84. J'ai l'impression que tout le monde puise dans la même banque d'images. En deux heures, j'ai vu une douzaine de fois le visage fermé de cette petite fille debout dans la foule. [In brackets my alternative translation].

26 'Finalement, on n'a pas eu ces scènes de débordement que certains journalistes (sûrement pas tous) ont appelées de leurs vœux.' Homel, p. 70.

27 Jonathan M. Katz, *The Big Truck That Went By: How the World Came to Save Haiti and Left Behind a Disaster* (New York, NY: Palgrave Macmillan, 2013), p. 278.

28 In her 2010 account, *Failles [Fault lines]*, Yannick Lahens comments in the following manner on the disempowering effect of this lexicon of resourcefulness and resilience: 'La resilience est devenu le terme commode, hâtif, souvent teinté d'exotisime, pour en parler, presque comme d'une essence. Le racisme n'est pas loin non plus.' 'Resilience has become a handy and quick term tinged with exoticism, to speak about it, and almost an essentialism. Racism is not far away either.' Yanick Lahens, *Failles* (Paris: Sabine Wespieser, 2010), p. 105.

29 Homel, p. 90; 'Ce n'était pas à cause d'un coup d'Etat, ni d'une de ces sanglantes histoires où vaudou et cannibalisme s'entremêlent – c'était un séisme. Un événement sur lequel on n'avait aucune prise. Pour une fois, notre malheur ne fut pas exotique. Ce qui nous arrive pourrait arriver partout.' Laferrière, p. 94.

30 Turner and others emphasise the need to go beyond the two standard models that have informed vulnerability analysis, namely the risk-hazard and the pressure-and-release models. Vulnerability, according to them,

> is registered not by exposure to hazards (perturbations and stresses) alone but also resides in the sensitivity and resilience of the system experiencing such hazards. This recognition requires revisions and enlargements in the basic design of vulnerability assessments, including the capacity to treat coupled human–environment systems and those linkages within and without the systems that affect their vulnerability. A vulnerability framework for the assessment of coupled human–environment systems is presented.
>
> Billie. L. Turner and others, 'A Framework for Vulnerability Analysis in Sustainability Science', *Proceedings of the National Academy of Sciences*, 100 (2003), 8074–9 (p. 8074)

31 Lizabeth Paravisini-Gebert, in her 2015 article on Marie Chauvet's *Fond des Nègres* 'as the most sustained exploration of deforestation' also turns to deforestation yet focuses, more specifically than Laferrière's city-oriented text, on

> the reality that without trees the topsoil will be washed away, and its consequence, that without topsoil (what the peasantry in the text understands as *the land*), the filial, cultural and religious rhizomes that give Haitian peasant culture its character cannot thrive or survive.
>
> Lizabeth Paravisini-Gebert, '"All Misfortune Comes from the Cut Trees": Marie Chauvet's Environmental Imagination', *Yale French Studies*, 128 (2015), 74–91 (pp. 74–5)

32 Homel, p. 119. 'Un tel jeu est difficilement compréhensible dans une ville aussi surpeuplée. Il exige trop d'espace pour un public restreint [...] Une minuscule balle blanche pour un si vaste terrain, ce qui semble une provocation de plus.' Laferrière, p. 125.

33 Homel, p. 119. '[i]ci la terre est bonne mais pas un arbre fruitier dessus'. Laferrière, p. 125.

34 Marquez Stathis, back cover.

35 'chronicle, n.', in *Oxford English Dictionary Online* (Oxford University Press, 2015) www.oed.com/view/Entry/32576?rskey=tBvYZE&result=1 [accessed 3 June 2015].

36 Marquez Stathis, back cover.

37 A quotation from Mary Catherine Bateson's *Composing a Life* (1989) opens Part I: '*she grasped the idea that one could study culture, one's own or that of others, truly attending to it rather than using the stance of an observer as a way to dominate*' [italics original]. Marquez Stathis, p. 1.

38 Jorge Luis Borges's 'Two English Poems' (1934):

> *I offer you that kernel of myself that I have saved,*
> *Somehow – the central heart that deals not*
> *in words, traffics not with dreams, and is*
> *untouched by time, by joy, by adversities.*

[Italics original] Marquez Stathis, p. 69

39 Marquez Stathis,back cover.

40 Marquez Stathis, p. 293.

41 Marquez Stathis, p. vii.

42 Marquez Stathis, pp. 180–1.

43 Ann Laura Stoler, 'Imperial Debris: Reflections on Ruins and Ruination', *Cultural Anthropology*, 23 (2008), 191–221 (p. 195).

44 'ruin, n.', in *Oxford English Dictionary Online* (Oxford University Press, 2013) www.oed.com/view/Entry/168689?rskey=rk5JU5&result=1&isAdvanced=false [accessed 12 March 2013].

45 'ruin, v.', in *Oxford English Dictionary Online* (Oxford University Press, 2013) www.oed.com/view/Entry/168690?rskey=rk5JU5&result=2&isAdvanced=false [accessed 12 March 2013].

46 Stoler, 'Imperial Debris', pp. 195–6.

47 Marquez Stathis, p. 70.

48 Beckett, 'Rethinking the Haitian Crisis', p. 33.

49 As the 767 transporting me through the clouds comes in for landing, I see white tents dotting the perimeter of the airport. This is my first glimpse of the "new Haiti". Once on the ground, we share runway space with a C-130 American military cargo plane and we park near a curiously named Planet Airways plane, harbingers that we have arrived in Disasterland.

Marquez Stathis, p. 80

50 'I feel as if I am standing in line amidst a group of foreigners waiting to return home from Disasterland', Marquez Stathis, p. 252.

51 Marquez Stathis, p. 219.

52 Marquez Stathis, p. 260.

53 Marquez Stathis, p. 219.

54 During her trip to Delmas, a neighbourhood of Port-au-Prince, the narrator observes: 'I'm not sure if it's just a matter of my eye needing to adjust to Haiti, or if the country has slipped even deeper into the muck. I'm afraid it's the latter.' Marquez Stathis, p. 180.

55 'muck', in *Oxford English Dictionary Online* (Oxford University Press, 2015) www.oed.com/view/Entry/123164?rskey=5XwKKH&result=1&isAdvanced=false [accessed 17 June 2015].

56 As Robert Muggah, in his analysis of the effects of the stabilisation mission in Haiti, suggests: 'Even before the notorious Duvalier dictatorships came to an end in the mid-1980s, Haiti was described alternately as "failing", "failed", and "fragile" in international policy circles.' Robert Muggah, 'The Perils of Changing Donor Priorities in Fragile States: The Case of Haiti', in *Exporting Good Governance: Temptations and Challenges in Canada s Aid Programme* ed. by Jennifer Welsh and Ngaire Woods (Waterloo: Centre for International Governance Innovation and Wilfred Laurier University Press, 2008), pp. 169–203 (p. 169). Beverly Bell also points to the prevalence of discourses of Haiti as a 'failed state':

> The message had long been drummed, but the tempo picked up after the earthquake, pounded by foreign governments, academics and media, such as *Time* magazine's piece 'The Failed State That Keeps Failing'. Later, Haiti would even come in fifth in the 2011 Failed States Index of the think tank Fund for Peace, after Somalia, Chad, Sudan, and the Democratic Republic of Congo.
>
> Bell, p. 80

For other examples of the operations of this discourse, see Tim Padgett, 'The Failed State That Keeps Failing: Quake-Ravaged Haiti Still without a Government', *Time Online*, 10 September 2011 http://world.time.com/2011/09/10/the-failed-state-that-keeps-failing-quake-ravaged-haiti-still-without-a-government [accessed 15 July 2013] and Tim Weiner, 'Life Is Hard and Short in Haiti's Bleak Villages', *New York Times Online*, 14 March 2004 www.nytimes.com/2004/03/14/world/life-is-hard-and-short-in-haiti-s-bleak-villages.html?pagewanted=all&src=pm [accessed 15 July 2013].

57 Marquez Stathis, p. 176.
58 'You are the light of the world. A city built on a hill cannot be hid.' Matthew 5.14.
59 John Winthrop, *A Modell of Christian Charity (1630)* (Collections of the Massachusetts Historical Society: Boston, 1838) 7, 31–48 https://history.hanover.edu/texts/winthmod.html [accesssed 01 October 2015].
60 Nick Nesbitt, *Universal Emancipation: The Haitian Revolution and the Radical Enlightenment* (Charlottesville, VA: University of Virginia Press, 2008).
61 J. Michael Dash, *Haiti and the United States: National Stereotypes and Literary Imagination* (Basingstoke: Macmillan, 1988), p. 136.
62 J. Michael Dash elaborates on this particular form of sea traffic:

> The first refugees who turned up on the shores of the United States were French families fleeing the violence of the Haitian War of Independence at the end of the eighteenth century. By 1915 the ships that find their way to Port-au-Prince are American warships that land Admiral Caperton and his Marines. The mass exodus of black, impoverished refugees from Duvalier's regime retraces the journey made almost two centuries earlier by white colonizers.
>
> Dash, pp. 135–6

63 Marquez Stathis, p. 88.
64 Marquez Stathis, p. 228.
65 The narrator is no more subtle in her interpretation of the former priest's presidency, equating it with the Duvalier dictatorship – both leaders are 'vermin'. She writes: 'The earthquake shook the foundation of the Haitian state, destroying Devil's House and creating a space for two vermin to crawl out of their hiding places from France and South Africa.' Marquez Stathis, p. 278.

66   The shell of a building still standing, like a silhouette rising from the hotel remains, bears a certain resemblance to the Acropolis in Greece. I soon find myself staring off into infinity. I hear the familiar echo of song rising from the hillside down below. It's comforting. The natural thought that develops is that I try to imagine what those final moments were like for Andrea and her colleagues in this building.

Marquez Stathis, p. 226

Acropolis (in 'ancient Greek ἀκρόπολις upper or higher city, citadel, especially that of Athens) is composed of two combining forms 'ἄκρο- acro-' (*OED*) and 'πόλις – polis-' (*OED*). 'Polis' denotes a type of town or a city and the former, similar to the 'Indo-European base as edge' and 'forming terms relating to height, or to the highest or foremost part of something'. 'acropolis, n.', in *Oxford English Dictionary Online* (Oxford University Press, 2015) www.oed.com/view/Entry/1854?redirectedFrom= acropolis [accessed 17 January 2015].

67 Marquez Stathis, p. 139.
68 The Trojan horse was a huge, hollow wooden horse constructed by the Greeks to gain entrance into Troy during the Trojan War. Greeks, pretending to desert the war, sailed to the nearby island of Tenedos, leaving behind Sinon, who persuaded the Trojans that the horse was an offering to Athena that would make Troy impregnable. Despite the warnings of Laocoon and Cassandra, the horse was taken inside. That night warriors emerged from it and opened the city's gates to the returned Greek army. The term 'Trojan horse' has come to refer to subversion introduced from the outside. See 'Trojan horse', in *Britannica Online* (2013) www.britannica.com/EBchecked/topic/606297/Trojan-horse [accessed 13 August 2013].
69 Marquez Stathis, p. 143.
70 Marquez Stathis, p. 145.
71 Genesis 22.1–24.
72 'sacrifice', in *Oxford English Dictionary Online* (Oxford University Press, 2012) www.oed.com/view/Entry/169571?rskey=4fvYPa&result=1&isAdvanced=false#eid [accessed 05 November 2012].
73 Marquez Stathis, p. 283.
74 Lilianne Fan, *Disaster as Opportunity? Building Back Better in Aceh, Myanmar and Haiti* (London: Humanitarian Policy Group, Overseas Development Institute, November 2013) www.odi.org.uk/publications/8007-resilience-build-back-better-bbb-aceh-tsunami-cyclone-myanmar-earthquake-haiti-disaster-recovery#downloads [accessed 13 February 2014], p. 1.
75 In fact, within the overall scale of loss, the casualties among international workers were relatively small – which is not to dismiss the narrator's acute sense of loss or her need to mourn the death of her friends and colleagues.
76 Popkin, 'Life in the Ruins', p. 106.
77 Homel, pp. 130–1.

> Le fait que les dictateurs l'on squatté, plus souvent qu'à leur tour, depuis plus de deux cents ans, ne le [Palais National] rend nullement indigne. Les gens n'ont jamais fait l'erreur de confondre le bâtiment avec son occupant. Ils espèrent un jour lui redonner sa splendeur. Si l'importance d'un édifice réside dans l'émotion que son absence déclenche, celui-là a une valeur plus que symbolique. Une houle d'émotion a submergé la ville, le pays même, quand on a su que le Palais avait explosé sous la violence de la première secousse.
>
> Laferrière, p. 137

78 Homel, pp. 63–4. 'Rien ne nous retient. Plus de prison, plus de cathédrale, plus de gouvernement, plus d'école, c'est vraiment le moment de tenter quelque chose. Ce moment ne reviendra pas. La révolution est possible, et je reste assis dans mon coin.' Laferrière, p. 66.

79   La radio annonce que le Palais national est cassé. Le bureau des taxes et contributions, détruit. Le palais de justice, détruit. Les magasins, par terre. Le système de communication, détruit. La cathédrale, détruite. Les prisonniers dehors. Pendant une nuit, ce fut la révolution.

> Laferrière, p. 29

'The radio announced that the Presidential Palace has been destroyed. The taxation and pension office, destroyed. The courthouse, destroyed. Stores, crumbled. The communication network, destroyed. Prisoners on the streets. For one night, the revolution had come.' Homel, p. 30.

80 Here the reference is both to the legal-political concept of 'state of exception', as elaborated by Giorgio Agamben, who builds on an earlier debate between Walter Benjamin and Carl Schmitt after Schmitt's publication of *Political Theology. Four Chapters on the Concept of Sovereignty* (1922), especially in relation to refugee camps and IDP-camps, as well as the conjunction of aid practices and international policies (e.g. Interim Haiti Recovery Commission led by Bill Clinton, the continued presence of MINUSTAH, protected by a mandate of immunity) which effectively replaced the Haitian state following the 2010 earthquake. In Agamben's discussion of the term, the state of exception 'is not a dictatorship, but a space devoid of law'. The state of exception defines 'a regime of the law within which the norm is valid but cannot be applied (since it has no force), and where acts that do not have the value of the law acquire the force of law'. Giorgio Agamben, 'The State of Exception' (Extract from a lecture given at the Centre Roland-Barthes (Universite Paris VII, Denis-Diderot) and an edited translation of 'Lo stato di eccezione come paradigma di governo': the first chapter of Agamben's *Stato di eccezione. Homo Sacer II* (Bollati Boringhieri, May 2003, Torino www.egs.edu/faculty/giorgio-agamben/articles/state-of-exception/ [accessed 13 August 2015].) State of exception is not something external but rather has to be understood as 'a threshold, or a zone of indifference, where inside and outside do not exclude each other but rather blur with each other'. Giorgio Agamben, *State of Exception*, trans. by Kevin Attell (London; Chicago, IL: University of Chicago Press, 2005), p. 23. This blurring is also visible in the involvement of foreign governments and stakeholders (e.g. OAS, UN) in the direct shaping of Haiti's internal politics (e.g. the 2010 presidential elections) and post-earthquake reconstruction policies.

At the same time, there are key limitations to Agamben's theorisations of 'the state of exception' and 'bare life' such as his extremely limited engagement with postcolonial spaces, as demonstrated in powerful critique by, within Haitian Studies, Kaiama L. Glover, Sibylle Fischer, and, more generally in Black Studies, by Alexander G Weheliye. See: Alexander G. Weheliye, *Habeas Viscus: Racializing Assemblages, Biopolitics, and Black Feminist Theories of the Human* (Durham: Duke University Press, 2014). Sibylle Fischer, among others, demonstrates the reductivism and 'dramatic abstractness' of the concept of 'bare life'. She writes:

> the abstract graphicness of the concept of 'bare life' and the lack of contextual detail in the theory that produces it, make it available for a highly ambiguous fantasy investment. We do not need to get entangled in complexities of historical roots and causes. We can speak of the political catastrophes of the present without getting caught in miserly pity and compassion, or a human rights discourse that ultimately testifies only to its own powerlessness. The extremes of violence of the contemporary state and the degradations imposed on vast populations can be invoked in the style of Greek tragedy, or the unflinching realism of those pictures my son had found on my desk. In Agamben's thought, 'bare life' is certainly not a figure of fantasy. Yet, the dramatic abstractness of the concept, which treats Auschwitz as the truth of Western politics, and its heightened rhetoric of life and death, of state of exception, of sovereign ban, and animalisation ultimately create an affective space where identifications and psychical enjoyment go unchecked. Agamben's 'impolitical politics' take place under the sign of death: the corpse, not in its universal inviolability, but infinitely violable. The problem is that violence separated from its roots and conditions, suffering pure and simple – even

when only referenced in philosophical terms – engages us in ways that no other subject does.

Sibylle Fischer, 'Haiti: Fantasies of Bare Life', *Small Axe*, 11 (2007), 1–15 (p. 8)

See also a rigorous discussion of the intellectual threads and omissions in Agamben's work and the concept of 'state of exception', among others, by Alessandra Benedicty-Kokken in Chapter One, 'Hegel and Agamben', and Chapter Two, 'States of Exception: Dayan, Trouillot, and Mbembe', in *Spirit Possession in French, Haitian, and Vodou Thought: An Intellectual History*, pp. 47–82; Alessandra Benedicty, 'Aesthetics of "Ex-centricity" and Considerations of "Poverty"', *Small Axe*, 16 (2012), 166–76; Kaiama. L. Glover, 'New Narratives of Haiti; or, How to Empathize with a Zombie', *Small Axe*, 16 (2012), 199–207.

81 Patrick Bellegarde-Smith, 'A Man-Made Disaster', p. 266.

82 'body politic', in *New Oxford American Dictionary*, ed. by Angus Stevenson and Christine A. Lindberg (Oxford University Press, 2011) www.oxfordreference.com. wam.leeds.ac.uk/view/10.1093/acref/9780195392883.001.0001/m_en_us1227580?rs key=sNSZri&result=4 [accessed 29 August 2013].

83 Beverly Bell, *Fault Lines: Views Across Haiti's Divide* (Ithaca, NY. and London: Cornell University Press, 2013), p. 84.

84 Homel, p. 55.

> Alors qu'il y a tout ce remue-ménage autour de Montana, juste à côté on demande de l'aide. […] Un homme, debout près de la voiture, fait remarquer, sans trop d'amertume, qu'il n'y a pas que le Montana. Mais c'est le lieu où on négocie de gros contrats et où d'importantes décisions politiques se prennent. L'hôtel favori des vedettes internationales qui s'intéressent à la misère des pauvres gens.
>
> Laferrière, p. 56

85 Disaster tourism, travelling to sites that have recently experienced disasters, is linked to forms of 'thanatourism' defined by Dann and Seaton as 'travel to a location wholly, or partially, motivated by the desire for actual or symbolic encounters with death, particularly, but not exclusively, violent death'. A. V. Seaton, 'Guided by the Dark: From Thanatopsis to Thanatourism', *International Journal of Heritage Studies*, 2 (1996), 234–44 (p. 240), in Anthony Carrigan, *Postcolonial Tourism: Literature, Culture, and Environment* (New York, NY: Routledge, 2011), p. 219. Recent examples of disaster tourism include bus tours in post-Katrina New Orleans, the Chernobyl Power Plant, the Exxon Valdez Oil Spill and many others. For example, a UK-based company, *Disaster Tourism*, offers bespoke disaster tours in the following categories: Tsunami Volunteering, Storm Chasing, Plane Crash, Volcano Disaster and Nuclear Disaster. See *Disaster Tourism* http://disastertourism.co.uk/disaster-tourism.html [accessed 13 March 2015].

86 Jonathan Katz in his *The Big Truck That Went By* (2013) traces the equally problematic 'relief career' of two American figures: Sean Penn and Bill Clinton.

87 Daniël J. Van Hoving and others, 'Haiti Disaster Tourism – A Medical Shame', *Prehospital and Disaster Medicine*, 25 (2010), 201–2 (p. 202).

88 Homel, p. 165. 'Le problème c'est que ces populations du tiers-monde ont développé, avec le temps, une vraie mentalité d'assistés. On sent bien qu'ils connaissent tous les rouages du système d'aide internationale. Ils l'ont étudié attentivement.' Laferrière, p. 174.

7 Future rebuilding and reconstructions

Through their portrayal of urban landscapes, *Rubble* and *Tout bouge* offer commentaries on Haiti's present devastation, processes of ruination that contributed to it, while also gesturing towards potential scenarios for Haiti's reconstruction, material as well as symbolic. These are coloured, respectively, for Marquez Stathis, by the more nostalgic sense of reassuring rural changelessness and urban unchangeability or, in Laferrière's collection, more hopeful emphasis on spatial continuity and the need to establish points of connection between the 'before' and 'after' of Haiti's capital. Here, too, the generic conventions embraced by the two narratives as well as the contrasting ways of translating the emotional impact of the event directly shape the texts' dissimilar visions of post-earthquake futures. Sharing the sense of impatience, irritation and urgency, the two texts sketch diverging paths towards a disaster-less future. Whereas *Rubble* roots its vision of Haiti's future in the turn to the recovered pastoral landscapes, *Tout bouge* imagines it from cracked, but reassembled, cityscapes.

Clearly overwhelmed by the sights of the devastated city, the narrator of *Rubble* struggles to envisage positive scenarios for the metaphorical and literal reconstruction of Haiti's capital, expressing a clear sense of frustration at seemingly surreal architectural renditions and projects of how the rebuilt city might look: 'They [the architects] might as well put up images of the Champs Elysee [*sic*] in Paris. For now, these renditions are utterly disconnected from the grim reality of Haiti's capital.'[1] The narrator's resentment is indicative of the dual attempt to mark a closure to the event while admitting the near-impossibility of doing so. At the same time, it points to the key challenge to any rebuilding efforts, namely the tension between radical ambitions for the city and the awareness, which at times gives in to disillusionment, of the difficulties involved in realising them. One such source of irritation, clearly discernible in Marquez Stathis's narrative and the tone of its descriptions of the city's reconstruction, is the sense of detachment some of these projects embody and the resulting practices of marginalisation, exclusion and external imposition of certain visions of development that they re-enact. The semi-ruined Palace, for example, is hidden away behind 'large panels depicting architectural renditions and planning announcements acting as a screen to reframe the crowd's vision'[2] – a telling illustration of wider politics of disenfranchisement. The public are thus denied access to the very site that purports to represent them provided instead with, at least for the narrator, disconnected and mismatched visions of a future Haiti.

Furthermore, the narrator's annoyance at the rift between the artist's vision and the actual context of the proposed site demonstrates the ways in which socio-economic privilege is mapped onto and reproduced in these urban reconstruction plans: 'It's a drawing that could be Lincoln Road, the Miami Beach pedestrian mall, with its vibrant cafes, bookstore, shops, and street life.'[3] Here, the symbolic significance of the image is more important than its accuracy. The Miami-inspired design of the new civic space appears to represent, like the golf course in *Tout bouge*, the ambition to be part of an economically privileged society. Additionally, it conjures up the imaginaries of the American Dream as well as the possibility of social and economic advancement of a community seen as a sum of individual efforts – a notion of progress that, paradoxically, is openly criticised in the text yet still underpins the narrator's vision for Junior's future.

This positive scenario she is trying to script for the boy, for example by making sure he is enrolled in a good private school and has English language classes, is a direct translation of this belief in self-advancement through hard work. At the same time, this hopeful vision is already undermined by the momentary recognition that this more future-affirming sequence of events is unavailable to most Haitians; nor does she seem to regard such a collective project as a realistic one. In effect, a clear tension emerges between, on the one hand, the narrator's willingness and belief in her ability to set up a structure that will allow Junior to navigate his way out of the debris and, on the other, her awareness, at times voiced in an embittered tone, of longstanding structural obstacles against which she is trying to fight, singlehandedly. The narrator's inability to resolve this tension translates, on one occasion, for example, into an impassioned dismissal of a passer-by's comment, whom she is trying to discourage from even contemplating a life abroad. Shortly after making the comparison between the architect's sketch and Miami, she observes:

> [a] young Haitian who is standing there [...] says irresistibly, 'It's nice there, isn't it?' I pause, wondering how best to offset the expectation that I have just unwittingly created. 'Yet, it's nice', I say. 'But like everywhere in the world, they have problems, too.'[4]

Here, the dynamics of space and place are made clear, with the American city embodying the young Haitian's desire to be part of it. The narrator's response reveals a partial acknowledgement of her own privilege as well as her inability, despite her repeated emphasis on affective and personal ties to the country, to negotiate and find her own position in Haiti and to fully engage with and respond to the complexity of the situation. In effect, the city, as experienced by Marquez Stathis, offers few, if any, realistic and feasible possibilities for a rebuilding of a collective in the aftermath of the earthquake. For her, the only way out of the debris, however narrow the path might be, is through individual efforts, out of the city and into the countryside.

Past perfect

Irritated and impatient with the imposed visions of post-earthquake reconstruction, yet equally eager to see and willing to help her friend, and the wider Haitian community, to rebuild after the earthquake, the narrator of *Rubble* turns away from the urban to rural sites and Haiti's natural landscapes as the more likely sources for the country's reconstruction and rejuvenation. Here, too, she tries to provide a judicious yet intimate account of care and concern, one which in its tone and the imagery of renewal, retreat and revival, seeks to somehow control the overwhelming sights of destruction and suggest the possibility for Haiti's more harmonious futures. In the process of doing so, the text reveals the difficulty of envisaging island environments beyond the binary opposition of seemingly tranquil, timeless and pre-modern, rural and the violently troubled urban, existing in an accelerated and jumbled time. At the same time, the text's struggle, evident throughout Marquez Stathis's account, productively and usefully reveals the necessity of confronting and accounting for Haiti's, as well as the wider Caribbean's, *globalised rural* – a shift that directly breaks away from the dehistoricising connotations of the 'sentimental pastoral'[5] imagery towards, after Graham Huggan and Helen Tiffin, post-pastoral imaginaries of Caribbean space and place. Away from oppositional framings of the countryside, in the negative mode, as anti-modern and non-urban and, in the positive mode, as a prelapsarian space, the Caribbean rural is at the heart of modernity formation, dynamics of consumption, migration, resource extraction,[6] and – as became clear during the days and months following the 2010 earthquake – disasters, their past, present and future.

After its opening personal remarks, *Rubble* shifts to offer background information about the narrator's relationship with Junior and to present the reader with an account of her stay in Haiti in the 1990s. It is during this trip that Marquez Stathis first frames her depiction of the country, which she retrospectively revisits in 2010, in terms of dualities, 'abundance and a lulling quality' versus the lack of 'economic opportunity',[7] and irreconcilable contrasts embodied in the environment:

> As I fell asleep, I remembered the explosive beauty of the flamboyant tree that captured my imagination […]. It was the first time I noticed the singular grace of this tree, and from then on, it would remain branded in my mind as a symbol of the beauty and pathos of Haiti.[8]

During her stay, the countryside had paradisiac qualities, functioning as an uncorrupted space that sustained her as she attempted to 'save' Haiti through her work for the UN. The use of the religiously rooted word 'grace' in the passage, followed by the pairing of 'beauty' and 'pathos', creates a vision of Haiti as a country outside time and history that, in binary terms, can only 'evoke sadness or sympathy'.[9] This mythologising image, which tellingly reveals the narrator's defamiliarisation and sense of estrangement from the surroundings she finds herself in, imparts *timelessness* to nature, suggesting that the narrator's experience of the natural world is one of transition from a complex and difficult present to a simple and comforting past, with the pastoral tone and imagery being used here to impart a sense of retreat and temporal displacement.

Having designated the Haitian countryside as a site of harmony and respite during her first stay, even following the January 2010 disaster the rural landscapes are repeatedly portrayed as a timeless and peaceful refuge that, if recovered on her journey, would seem to indicate the country's ability to rebound; that is to bounce back into a future that, at once, is connected to and preserves the past:

> I am on a quest today to find trace of pastoral Haiti – bright sugarcane fields, neatly tended rows of crops, children attending village schoolhouses, women riding horseback to market, people living simply off the land, just as they did hundreds of years ago.[10]

Here and elsewhere, the natural beauty of the tropics, that she hopes to rediscover, seems to counter the chaos of post-earthquake devastation while also determining the narrator's own emotional equilibrium and her belief in Haiti's ability to recover in the aftermath of the January earthquake. The landscape has almost Edenic qualities of plenitude. It functions as a haven and its beauty compensates for the violence and chaos surrounding this seemingly secluded and uncontaminated space which offers 'a retreat from politics into an apparently aesthetic landscape that is devoid of conflict and tension'.[11] In effect, the countryside becomes for Marquez Stathis a space of reconciliation and hopeful unity for which she yearns but struggles to see elsewhere in post-earthquake Haiti.

Consequently, the harmony-seeking language of the passage, which reflects the narrator's own attempt to overcome the emotional shock at the size and scale of destruction, risks revisiting standard tropes of the timeless, the exotic and the pre-modern and suggests a sense of prelapsarian idealism, political quietism and innocence, implying a naïve and simplistic, if not retrograde, vision and politics of a given aesthetic representation.[12] Contrary to the narrator's hopeful and comforting impressions, violence has never been absent from this imaginary garden. The unmentioned extermination of the native Taíno population, plantation slavery and deforestation all violently shaped and continue to shape this seemingly bucolic landscape.[13] These uncritical diction and imagery of the passage, coupled with romanticised evocations of an undefined harmonious historical past, in effect, greatly, if not entirely, undermine the text's earlier critique of the ethnographic gaze. They reveal how comfort-seeking reaffirmations of disaster-free space, understandable attempts to re-gain individual emotional balance, too risk being saturated with colonial discourse, rooted, to a great extent, in romantic histories of 'the Noble Savage' and mystical 'native space'.

Such sentimental figurations of the tropics were directly linked to the history of colonial conquests and, at different historical moments, have been presented through tropes of wildness and savagery as well as images of prelapsarian innocence and purity. At first, the tropics evoked horror and disgust; only later would the New World evolve 'into a utopian site – a positive, idealized metamorphosis'.[14] This positive reimagining of the tropics,[15] a route to the sensual sublime,[16] influenced by socio-political changes in Europe, was accompanied by the introduction of the idea of a 'Noble Savage' at a time when the Old World was witnessing an increasingly critical attitude towards the nobility and their privileged status, with

> the idea of the Noble Savage [being] used, not to dignify the native but rather to undermine the idea of nobility itself [and to] represent not so much an elevation of the idea of the native as a demotion of the idea of nobility.[17]

Michael J. Dash, building on Hayden White's analysis, stresses that the trope of the Noble Savage served the ideological needs of a rising bourgeoisie demanding revolutionary change, 'for it at once undermined the nobility's claim to a special human status and extended that status to the whole of humanity'.[18] White's and Dash's remarks hint at the duality of those literary representations of tropical nature which served to critique European society while re-asserting an essentialist notion of otherness and, with it, the related concepts of the 'primitive' and the 'exotic' and the growing association of sensuality with the tropics. These tropical forms of otherness, however, were not only evoking places in need of being 'saved', but were also vehicles of 'salvation' for their creators, allowing them to escape – albeit only imaginatively – from the constraints of societies hidebound by puritanical values.[19] Echoing this desire to be saved from the sites of overpowering devastation and suffering, Marquez Stathis's reading of Haiti's pre- and post-earthquake rural landscapes throughout her travel-memoir is overwhelmingly defined by a sense of nostalgic longing for an impossible return to, and equally impossible repetition of, an idealised past that functions as an antithesis to the catastrophic present.

However, far from the sense of tranquillity and simplicity that the term seems to evoke and that Marquez Statis's depictions conjure, the wider category of the pastoral is anything but stable, and its divergent uses, particularly in relation to different geographical, historical and cultural contexts and literary traditions, suggest a confusing versatility. Attributed to 'almost bewildering variety of works',[20] the term pastoral is, increasingly, used to denote the content of a given work, 'ranging from anything rural, to any form of retreat, to whatever form of simplification or idealization',[21] rather than 'the artifice of [a] specific literary form'.[22] Attempting to trace these shifts, Terry Gifford identifies three main ways in which the pastoral has been employed. First, as a historical literary category with its roots in the Greco-Roman tradition and the idylls of Theocritus. Second, as a qualification of content that refers to a specific, yet in no way homogenous, literary mode for representing countryside life and rural landscapes, 'with an implicit or explicit contrast to the urban',[23] that has undergone numerous modifications, each shaped by the particular socio-political and cultural context of its period,[24] and has influenced a large number of other literary genres: drama, the elegy, the epic and, more recently, the novel.[25] It is the third use of the pastoral as a pejorative qualifier which draws on this sense of prelapsarian idealism, political quietism and innocence and often implies a naïve and simplistic, if not retrograde, vision and politics of a given aesthetic representation, that has been widely picked up in contemporary readings of the genre. When translated to the various socio-political contexts with which it engages, the pastoral's pejorative characteristics can easily, as Annabel Patterson explains, 'emphasise the stability, or work towards the stabilisation, of the dominant order, in part through the symbolic management – which sometimes means the silencing – of less privileged social groups'.[26]

Yet, the pastoral's oft critiqued potential to reaffirm lost innocence and nostalgia can serve in contrasting, sometimes contradictory, ways. Pastoral, for example, can be conscripted, in a reverse mode, for critique by challenging complacent visions of political escapism and retreat into an idealised landscape.[27] The 'ideological grammar' of the pastoral is 'always contingent, always shifting':[28] it is more than just a retrograde form rooted in bourgeois ideals.[29] Similarly, for Peter Marinelli, pastoral is not inherently evasive of social problems but has the potential to stage confrontation. The poetic persona in search of Arcadia can never fully escape the city and its complexity. Thus, to arrive in Arcadia is only 'to have one's problems sharpened by seeing them magnified in a new context of simplicity',[30] and a temporary state can only be a 'prelude to return'.[31] It remains unclear in this formulation, as well as in *Rubble*, however, whether this realisation leads to a more critical view of the past that goes beyond sentiments of nostalgia and loss.

Whereas Gifford's analysis helpfully charts the many trajectories of the pastoral in order to underline its conceptual relevance, recent criticism has introduced the sub-categories of 'anti-pastoral' and 'post-pastoral' in order to offer a more specific and contextually-nuanced consideration of the significance of pastoral as a literary-historical mode for contemporary debates on narrative aesthetics and ethics. The former sub-category can be 'characterised as the opposite of pastoral in that [it] entail[s] a journey to a kind of underworld and return harrowed rather than renewed'.[32] The latter, a rejection of an easy literary *cliché*, is marked by respect and 'an awe in attention to the natural world [which derive] not just from a naturalist's intimate knowledge or a modern ecologist's observation of the dynamics of relationships, but from a deep sense of the immanence in all natural things'.[33] In this last redefinition, post-pastoral signifies a literary mode 'informed by ecological principles of uneven interconnectedness'.[34]

Characterised by 'educated understandings of the symbiotic link between environmental and social justice, at both the local level and beyond',[35] post-pastoral, as Huggan and Tiffin demonstrate, allows, in effect, for a more nuanced engagement with the landscape's socio-material histories, emphasising a sense of interconnectedness and embeddedness that is crucial to an understanding of postcolonial Caribbean ecologies, in general, and the 2010 Haitian earthquake, in particular:

Pastoral, in this sense, is about the legitimation of highly codified relations between socially differentiated people: relations mediated, but also mystified, by supposedly universal attitudes to land.[36]

This last point echoes some of the earlier criticisms of the pastoral mode, putting, however, more emphasis on the relational and processual aspects of landscape formations and the hierarchies and power relations they embody. Where Huggan and Tiffin turn to the post-pastoral in order to stress the sense of interconnectedness and historicity of the landscape, Beverly Bell employs pastoral imagery and botanical metaphors specifically in order to challenge the depoliticisation of Haitian nature and put forward a contextualised engagement with Haiti's natural landscapes as complex environments, including 'all of the external abiotic and biotic factors, conditions, and influences that affect the life, development, and survival of an organism or a community'.[37] In effect, she calls for more empowering models of individual/collective response to testing events such as the 2010 earthquake. This broader definition, central to the aforementioned category of post-pastoral, stresses the interconnectedness of abiotic and biotic components and allows for a conceptualisation of social and ecological resilience, away from connotations as a passive, or inherent, capacity towards notions of active response and positioning.

Overturning the discourses on Haitians' passive resilience, Beverly uses pastoral imagery and proverbs to argue that resilience is not a submissive state of waiting and inertia, but rather the manifestation of an active and informed refusal to succumb to the violence of a particular event:

The bamboo symbolizes the Haitian people [...]. The bamboo is really weak, but when the wind comes, it bends, but it doesn't break. Bamboo takes whatever adversity comes along [...] that's what resistance is for us Haitians; we might get bent [...] but we're able to straighten up and stand.[38]

In her use of pastoral, Bell demonstrates the potential of the mode to expose the violence inherent in any transformation of Haiti's landscapes. Finally, within this dual movement of challenging and reversing naturalising discourses of resilience, Bell's imagery also makes reference to a popular Haitian proverb, 'Ayiti se bwa wozo, li kap pliye men li pap kaze' ('Haiti is like reed' (or 'bamboo'). 'It bends but it doesn't break')[39] which asserts individual and collective resolve to confront the difficulties the country might be facing, whether internally or externally.

What becomes clear, across these diverse reworkings of the pastoral mode, is that categories of 'nature' and 'landscape' have a two-fold spatial and temporal character: they denote a particular space, but also the ongoing processes that shape it. As such, they are entwined within a net of uneven relations and interconnected vulnerabilities linking socio-economic inequality, dispossession, and exploitation of land and labour, including 'imperial genocide and colonial plantation slavery',[40] deforestation, tourism and, not least, disasters, all of which influence the 'natural' transformations and reconfigurations of space and landscape in the Caribbean and, more specifically, Haiti.[41]

Future perfect

One of the few envisaged scenarios for a wider-reaching community recovery rather than individual-focused post-disaster development, Marquez Stathis's text is indeed the growth of 'sustainable tourism' – effectively a translation of the pastoral vision into the corporate dynamics of the neoliberal market economy. Again, attempting to translate her emotional investment in Haiti into a language of seemingly impartial observation, the narrator enthusiastically commends her friend's 'unique' contribution to Haiti's growth:

> […] Carole is part of a unique effort to discover and preserve Haiti's caves, unique natural underworlds that retain traces of the island's pre-Columbian Taino culture, in hopes of renewing tourism to Haiti in a way that is both good for the economy and the environment. The Caves of Haiti project, of which she is a co-founder, has garnered support from UNESCO.[42]

While this vision is consistent with the text's emphasis on the preservation of Haitian unspoilt countryside, considered primarily as a rural idyll, it also raises questions as to ecotourism's links with other, more obviously opportunistic forms of tourism, e.g. disaster tourism and 'disaster business'.[43] Tourism, rather than assisting the local population or protecting the environment, can potentially lead to even greater social stratification, creating a class of poorly paid workers, increasing the exploitation of limited environmental resources and contributing to a growing reliance on external investment.[44] Michel Martelly's[45] infamous declaration – made on 28 November 2011 at the inauguration of the Caracol Industrial Park (CIP) in the north of the country – that 'Haiti is [now] open for business',[46] epitomises this neoliberal model of development.[47] Before the construction of the CIP, the Caracol municipality in the north-east of Haiti was the so-called 'breadbasket of the Northeast department', yet from the start of construction, farmers were expelled from their land and, contrary to previous claims of tens of thousands of jobs, the site employs approximately 1,388.[48] The future return envisaged here is not to the prelapsarian garden that Marquez Stathis conjures but rather to the rebuilt sweatshop – the countryside's dominating motif. If *Rubble* does not necessarily subscribe to this model, then it does little to consider other potential routes for collective change and post-earthquake reconstruction outside of further commodification of fragile ecosystems – already strained by expanding urbanisation and the little-controlled yet highly-encouraged and growing tourism industry – that leads to an increased vulnerability of Haiti as well as the wider Caribbean.

Within this linear, at once nostalgic and progress-oriented, vision, the earthquake rubble is quickly overgrown by lush greenery; the debris safely out of sight. This emphasis on the recovery of the pastoral Haiti, as a physical space as well as a certain imaginary of it, joins ecological renewal with restoration of a sense of peace and tranquillity, one which the narrator seeks for herself and which those affected by the disaster need even more. Translation of the narrator's empathy and her own distress and conceptualisation of rubble and ruins as loss and absence, this overly hopeful pastoral vision, however, is equally a reiteration of the false binary between Haiti's rural and urban sites – a positioning that falls short of accounting for the entanglement between them and the resulting need for an integrated vision of reconstruction, not just building over or casting away the earthquake rubble. Consequently, Marquez Stathis's account straddles, and ultimately struggles, to reconcile this narrative tension between a hopeful turn to bucolic imaginary – of a pre-modern, pre-disaster past – her sense of affective-writerly commitment to Junior's and Haiti's future, and her memoir's mixed character as a social commentary and an emotional chronicle of an individual journey of recovery. Although the narrator's pastoral reflections on Haiti's past landscapes are in line with her writerly method of emotional honesty, they effectively preclude an analytical consideration of the *historicity* of Haiti's rural landscapes that is characteristic of post-pastoral aesthetics (at least in Huggan and Tiffin's sense). Whereas, on the personal level, the pastoral imagery of the memoir can still effectively function to reflect the narrator's own quest for balance and emotional stability, as a metaphor for collective experience of the disaster, it offers no such reaffirming vision. Rather, the backward glance towards the coloured past works as a mode of political escapism, which effectively negates the complex and violent processes that once defined Haiti, while also upholding the sentimental view the narrator longs for – that of an untroubled retreat from the chaos.

Such depictions also fetishise the idea of 'wilderness', equating it with 'nature' as a category of landscape, ignoring that there 'is no singular "nature" as such, only a diversity of contested natures [...] constituted through a variety of socio-cultural processes'.[49] These uncritically rehearsed Edenic evocations point back to pastoral's oft critiqued ideological underpinnings, which also echo in the ultimately dehumanising discourses of Haiti's 'uniqueness', and are questionable in the context of the Caribbean, a geographical region defined by 'imperial genocide and colonial plantation slavery',[50] as well as ongoing exposure to multiple environmental risks that continue to shape island ecologies to this day. As a result, against the author's ambition to counter stereotypical portrayals of Haiti, rural spaces are still timeless and, likewise, the population that inhabits this imaginary Eden is detached from any material contexts that might surround it. In effect, the text does little other than reiterate tired metaphors that place Haiti outside global ecologies and portray it instead through a gesture of depoliticising distancing, as an enchanted far-away place of 'Voodoo, Zombies, and Mermaids'.[51]

Finally, this vision of restoration duplicates, in the ecological realm, the narrator's idea of post-disaster social transformation with an individual's efforts, whether that of her environmentally-minded friend or of Junior, having a casual, always positive effect on the wider society. Minimising structural constraints and sense of individual/collective and ecological interconnectedness, this vision allows the narrator to give meaning to her involvement in Haiti, in Junior's life, and see it as necessary commitment rather than intervention, inscribing it in a narrative of a possible socio-ecological recovery – one triggered and guided by an external yet well-willing helping hand. In effect, a narrative of emotional journey, *Rubble* emerges as a narrative *pledge*: an expression of commitment and a promise, made to Junior, of a better future at hand.

Future continuous

Whereas Marquez Stathis's 'emotional chronicle' looks to the idealised rural past and locates her hopes in individual efforts in an attempt to suggest some forms of post-earthquake rebuilding and extension of the rural, with the mixed generic conventions and emotive diction limiting the critical insights her narrative can offer, Laferrière's more detached narrative partly departs from its guiding principle of non-linear organisation in its portrayal of the country's reconstruction. In contrast to his earlier, more radical framing of the city's devastation, the narrator insists on the importance of linking pre-earthquake histories of space and place to possible post-disaster futures, and the text's customary focus on fragmentation is replaced by a collective call to look beyond and above the current destruction. At the same time, this emphasis on connectedness to the past is unlike Marquez Stathis's vision of individually-focused advancement and rush out of the rubble. Rather, in the face of loss and void, the narrator takes up his position – *prendre position* in French both to take up one's position and take a stance (*Reverso*) – among the debris, among and with those who are there.

While still obviously concerned with and overwhelmed by the scale of the material devastation, the narrator pauses for a moment, to observe, to listen and to make a tentative sketch of the post-disaster 'human landscape' which, for him, is the only possible source of continuity between the past and the present. From the long-shots of what is no longer there, the narrator zooms in on who *is* and who *will be* there, dwelling in and remapping the radically altered topography of Port-au-Prince:

> The earthquake didn't destroy Port-au-Prince; no one can build a new city without thinking of the old. The human landscape counts. Its memory will link the old and the new. Nothing is ever begun from scratch. It's impossible, in any case. All we do is continue. There are things you can never eliminate from a trajectory, like human sweat. What should be done with the two centuries, and all they contain, that preceded Year Zero? Throw it all in the garbage? A culture that pays attention only to the living risks its own death.[52]

In this extended quotation, the narrator directly challenges the temporal logic behind the creation of a post-earthquake 'year zero', which would place the event outside of the multiple contexts that inform it while equating the whole of Haiti with the destructed capital. The 'year zero' motif risks discarding the country's earlier history with the force of the event and its impact configured as a moment of complete erasure. This direct commentary on such temporal bracketing also echoes Saint-Éloi's attempts to forge new History in the aftermath of the earthquake (Part II). Whereas *Haïti, kenbe la!* works with ideas of reverberation and the vision of the country's history as a series of tremors, resounding shocks that modify and amplify the volume of subsequent events, *Tout bouge* contrasts the images of the fractured city with the ideas of ruptured historical time.

Differing in their narrative aesthetic, both *Tout bouge* and *Haïti, kenbe la!*, however, join their recognition of the earthquake as an unprecedented break, on the personal level, with an emphasis on the link between the event and its histories – whether understood in positive or negative terms. The earthquake, in other words, does not annihilate previous histories and memories of Port-au-Prince, but rather adds further layers of complexity to its urban topography. The urgent yet necessarily slow remaking of a life in this post-disaster city will be akin to the reconstruction of the broken vase – a poetic image used by Derek Walcott in his 1992 Nobel Prize lecture – that symbolises the Caribbean. For the St Lucian poet, the region's present is inseparable from its past. This knot of ruptured histories, redrawn, cut short and reassembled in the 'new world' is, for Walcott, 'the basis of the Antillean experience, this shipwreck of fragments'.[53] Resounding through Laferrière's collection, this notion of a splintered past is central to the narrator's attempt to dwell among the rubble and to see a life and new lives beyond it. It follows that the reconstruction of this symbolic vase, here as Haiti, will be akin to, in Walcott's gloss, a difficult gathering of the 'disparate, ill-fitting' pieces, 'our shattered histories, our shards of vocabulary [...] whose restoration shows its white scars'.[54]

Moreover, the post-earthquake destruction, forcibly etched onto the fabric of the capital, reshapes Port-au-Prince, while spatial coordinates are reorganised in relation to the city's former buildings and collective sites. As a result, a hybrid topography emerges, combining old reference points with new structural features:

> That situation created a new reality and people had to adapt fast. 'You know where the Caribbean Market used to be? Well, you go past it, and then two buildings that collapsed ...'. To the landscape of this crumbled city, people have added elements of the old one still present in their memory.[55]

Port-au-Prince might indeed be in ruins, yet even these destroyed spaces are not devoid of meaning and history. On the contrary, these scars and 'cracked heirlooms'[56] are given profound significance, evoking at once the pre-earthquake capital, the moment the disaster struck, and the resulting losses, as well as pointing to possible future reconstructions of the city. As the narrator makes clear, neither the event nor its aftermath can be divorced from the histories that underpin rural and urban space; it follows, then, that any rebuilding of the capital is itself a process of continuation, redefinition and negotiation – however different the new city might be. In refusing to erase his memories of Port-au-Prince and its architecture, the narrator points to the risk of seeing post-disaster reconstruction in terms of a return to an imagined past or a restoration of a pre-disaster state or, in an equally limiting mode, as a complete break with the past. Rather, drawing inspiration from the multi-layered city space, post-earthquake rebuilding should be conceived in terms of remaking and *renewal*, that is as a daring improvement rooted in, but not confined to, collective histories and memories of space.

Two separate entries, 'Une nouvelle ville' ('A New City') and 'Une ville d'art' ('A City of Art'), suggest some ways of translating these integrated notions of urban reconstruction into practice. The two sketches foreground the relationship between the city's topography and social stratification, underlining the importance of working against current hierarchies of space and exclusion. At the same time, the entries call for a creative reclaiming of the city so that it might evoke associations other than those of chaos, violence and vulnerability:

> Why not consider painting certain neighborhoods? Or turning Port-au-Prince into a city of art where music could play a role too? Haiti should use this truce to change its image. We won't have a chance like this a second time, if I can put it that way. Let's show a more relaxed face.[57]

Capturing a moment of radical optimism, grounded in the narrator's willingness to rethink what post-disaster reconstruction might mean even if his own diasporic position complicates his contribution to this renewal, the passage does not aim to glorify the city's poverty. Rather it hopes to reposition Haiti and its people so that they are no longer reduced to the category of 'the poorest nation in the Western hemisphere'. Still, the idea of repainting neighbourhoods might seem extravagant, or even unethical, at a time when even the more basic forms of reconstruction have not yet been completed. However, since the first French edition of Laferrière's text, the Haitian government carried out a similar, rather controversial, initiative to repaint Jalousie,[58] one of the capital's informal neighbourhoods, home to over 40,000 people, which borders with the affluent Pétionville district. Allegedly inspired by the work of Préfète Duffaut (a famous Haitian painter who passed away in 2012), the project received a mixed response from both inhabitants of the neighbourhood and foreign commentators who saw it as further contributing to 'poorism' (poverty tourism) or slum tourism,[59] not to the long-term well-being of the local community. Despite the controversies, that was not the last project which combined art, reconstruction and alleged development.[60] In 2015, Haas&Hahn of Favela Painting were invited by Cordaid (a Dutch NGO with a long-term presence in Haiti) to organise a painting project in Port-au-Prince, a centrally-located Vila Rosa neighbourhood,[61] that would create an opportunity for 'applying an artistic approach to CMDRR (Community Managed Disaster Risk Reduction)'.[62] The two projects' bold aspirations seem to translate some of Laferrière's urge to envisage radical remapping of the city space. Yet, if not accompanied by long-term ordinary improvements of the community's daily life, the explosion of colour only risks camouflaging 'the violence that produces "bare life" '.[63]

Furthermore, the reconstruction process as envisaged in *Tout bouge* attempts to connect the past, present and future visions of the capital, by repeatedly emphasising the non-material dimensions of cityscapes: where previous entries contemplated the 'human landscape', here the narrator speaks directly of the spiritual significance of space. But the text in no way suggests that sacred space is timeless or it can only be found in the countryside; rather, he wants to forge connections between the reconfiguration of urban landscape and the recreation of social identity, hoping that a more inclusive urban topography can be drawn in the aftermath of the disaster:

> People have [Everyone has] the right to say what kind of city they want to live in. Even better, they should be able to get involved in drawing up the plans.... The most important material is the spirit. A spirit open to the world, not concentrated on itself. [...] A new city that will compel us to enter a new life. That's what will take time. Time we haven't given ourselves.[64]

Here, this outward-looking 'spirit of the city' stands above all for the social transformation that the new space might trigger and for the collective dimension of this long-term and permanent change that the narrator hopes for: a new city, 'une nouvelle ville', will force us to start a new life, 'a nouvelle vie'. In contrast to the more distanced and critical remarks on aid and agency in Haiti in previous entries, here the narrator's voice identifies with the 'we' of the affected community, seeking a new life in a new city. Indeed, as he admits elsewhere, the capital and the lives of its inhabitants will never be the same: the challenge then, for the narrator, is to draw on this experience of rupture and a sense of being stuck in a halting impasse in ways that still allow him and the readers to imagine a life-affirming yet far from naïve future.

A radically redefined space, yet one that is connected to pre-disaster pasts, the post-earthquake city emerges as a crossroads, *kalfou* (in Haitian Kreyòl), a meeting place and a coming together of the old and the new. It is at once a juncture and a symbolic turning point as there can be no simple going back to the pre-earthquake Port-au-Prince. Haiti's capital is, then, best seen as a palimpsest, with much of the original writing superimposed or altered but still retaining traces of its earlier form (*OED*).[65] Equally, urban reconstruction, however boldly imagined in a moment of daring projection, adds to that layering and rather than just being rooted in ideas of complete rupture and a new beginning, 'year zero', commences with the lengthy and laboursome removal of debris – an essential first step that clears the way for any perspective reworkings of the city space. Still, the narrator, himself struggling to define his position within or outside of the national community and the wider group of those directly affected by the earthquake, is aware that such profound redefinitions of city space require time, which no one can afford.

Conclusion

Differing widely in temporal scope and narrative perspective, the two literary responses analysed in this chapter offer equally divergent interpretations of ruination, reconstruction and remaking in the aftermath of the January 2010 earthquake. At the same time, both texts chart a future-oriented vision, attempting to envisage the possibility of a disaster-free life from or away from the ruins. In the process of doing so, both narratives expose the discursive and conceptual limitations of antithetical notions of reconstruction considered as a hopeful extension into the future of a complete and completed past – a move from past perfect to future perfect – or as a complete erasure and the sweeping aside of the past and what still remains of it.

In its evaluation of post-earthquake recovery, *Rubble* extends the metaphor of devastated urban space to suggest the near-impossibility of rebuilding the capital of Haiti: post-earthquake city ruination seems destined to continue into an indefinite future. In this and other ways, the narrative juxtaposes urban *unchangeability*, the perceived impossibility of any real change occurring in Port-au-Prince in the future, with the nostalgic sense of *changelessness* embodied in the Haitian countryside. Future regeneration, in effect, is envisaged primarily in non-urban terms as a retreat to this recovered bucolic garden, an emotional as much as a physical space, resulting in an ahistorical presentation of nature as, at once, a refuge of the past and a resource for the future. Consequently, the narrative roots its vision of collective recovery in unproblematised images of a countryside rural – now reframed within the paradoxical logic of profitable yet always pristine landscapes – and a highly individualised narrative of social ascension, the only possible way of navigating out of the maze of urban rubble. In effect, *Rubble*, a narrative pledge of commitment, productively reveals the difficulties of reconciling the sense of urgency of post-disaster rebuilding, its teleological and linear logic of acceleration and leap into a rebuilt future, with the open-ended, necessarily slow and asynchronous individual/collective efforts to reconstruct one's life. As such, the travelogue also clearly demonstrates how the rhetoric of 'reconstruction' as a simple return or a complete break can work to erase or repeat, in the recursive mode, the history of violent 'cultural remaking[s] of [Caribbean] human, floral, and faunal populations'.[66]

Moreover, the linear form of Marquez Stathis's account, in which the earthquake is just one of many experiences that shape her friendship with Junior, suggests the possibility of achieving a positive 'resolution' of the disaster, illustrating the narrator's yearning for a sense of closure, on personal level, as well as her hope to contribute directly to Junior's future in the aftermath of the disaster. Additionally, through its mixed form, which relies on 'emotional honesty' as justifying grounds for sometimes crudely general socio-cultural observations, *Rubble* hopes to create a 'complete' view of the January disaster in which personal reminiscence doubles as a source of analytical objectivity. This possibility, however, soon dissolves; and through the tensions that emerge, the text points to the key issue of post-earthquake literary aesthetics: the problem of how to account for the event's highly particular character, its unprecedented emotional, material and symbolic impact, and the distinct context that informs it without falling back into discourses of 'uniqueness' regarding Haiti's present and future as well as its past.

Responding directly to this challenge through its kaleidoscopic-like structure, *Tout bouge* suggests the impossibility of achieving a 'total' or synoptic view of the disaster that Marquez Stathis attempts to chart in her narrative. In formal terms, instead of employing narrative focalisation as a way of circumscribing the aftermath of the earthquake – an attempt to delimit the scale of devastation and control the psychological impact of the event – *Tout bouge* continually questions the narrator's ability to come to terms with and find an adequate literary vehicle to express the sheer scale of devastation and loss. Each entry focuses on a different object, an encountered site or person, and in this way reveals interconnections between the personal, environmental and socio-political histories behind the January 2010 tremors. Laferrière's collection of short essays and sketches does not provide one simple answer or a solution, neither does it resolve the tension between the narrator's diasporic position, his shifting sense of belonging to the national community and his vision as well as potential contribution to future reconstruction. Instead, the collection repeatedly asserts the importance of the *interim*: the in-between, a moment of pause with those left behind, who become a bridge between the city's past, its crossroad present and future.

This vision of Port-au-Prince as *kalfou* equally emphasises the splintered *histories* of space, the ways in which rural and urban landscapes belong to a complex web of global interrelations and modernity formation. Consequently, the text's vision of reconstruction, unlike the initial depictions of the revolutionary breakdown of things, suggests a palimpsestic view of the post-earthquake capital as a radically altered space, yet one that builds upon rather than erases the city's earlier fabric. The layered, potentially more hopeful, perspective is equally a recognition that the time needed for this joint process of rebuilding the city and reconstructing lives in the aftermath of the earthquake is unavailable to most. In effect, the palimpsestic vision of the city space reveals the continuity, in its negative mode as persistence, of the very dynamics and structures of power, exclusion and inequality that defined the topography of Port-au-Prince and that have not been fully erased.

In sum, as they draw to a close, each of the two narratives reveal, without entirely resolving, the tension between the divergent temporality of rebuilding, reconstruction and remaking and the real time available to most. While attempting to conjure a more affirmative vision of urban and rural present and future, these narratives expose respective limitations of diction and of narrative form; both the hyperbolic linear account and the itemised, non-linear and detached collection cannot confine nor fully order the polyphonies and cacophonies of *goudougoudou*. This challenge of discordant temporalities of 'moving on' – the joint urge and impossibility to do so – is equally evident in the next three texts, to which I now turn, which contemplate the possible and hoped for transformations of the self and the collective in the aftermath of the event. The January 2010 tremors in Haiti certainly produced a radical reconfiguration of space and self, but they also took place within a layered discursive, historical and ecological continuum – which cannot be confined to a day or a single place and continues to have its own 'natural' history today – that, as Woolley's, Lake's and Wagner's narratives demonstrate, is an integral part of the oft contrasting personal and collective meaning-making frameworks and cosmologies of crisis, rescue and recovery.

Notes

1 Marquez Stathis, p. 228.
2 Marquez Stathis, p. 228.
3 Marquez Stathis, p. 228.
4 Marquez Stathis, p. 228.
5 Here, I draw on Leo Marx's use of the term. Marx, in his celebrated *The Machine in the Garden: Technology and the Pastoral Ideal in America* (1964), distinguishes between 'the complex and sentimental kinds of pastoralism' (p. 25). The latter, visible throughout Marquez Stathis's text, is a widely diffused popular and sentimental type of pastoralism and is 'an expression less of thought than of feeling' which expresses itself in 'the mawkish taste for retreat into the primitive and rural felicity' (pp. 5–6). Leo Marx, *The Machine in the Garden: Technology and the Pastoral Ideal in America* (Oxford: Oxford University Press, 2000).
6 Mimi Sheller, *Aluminum Dreams: The Making of Light Modernity* (Cambridge, MA. and London: MIT Press, 2014).
7 Marquez Stathis, p. 262.
8 Marquez Stathis, p. 24.
9 'pathos, n.', in *Oxford English Dictionary Online* (Oxford University Press, 2013) www.oed.com/view/Entry/138808?redirectedFrom=pathos [accessed 1 August 2013].
10 Marquez Stathis, p. 159.
11 Terry Gifford, *Pastoral* (London and New York, NY: Routledge, 1999), p. 11.
12 Roger Sales, for example, in his *English Literature in History, 1780–1830: Pastoral and Politics* (London: Hutchinson, 1983), mounts a sharp attack on the pastoral mode which, in his political reading of it, represents '[r]efuge, reflection, rescue and requiem'. He characterises the pastoral as an escapist mode since it seeks refuge in the country and offers a simplified vision of country life in which the values of the past are rescued by the text, creating a falsified reconstruction of a much more complex reality. For Sales, the pastoral mode was used, in effect, to prevent any questioning of the extant social order. Roger Sales, *English Literature in History, 1780–1830: Pastoral and Politics* (London: Hutchinson, 1983), p. 17.
13 As Henry Paget makes clear, 'the plantation became the institution that provided the broad framework through which subsistence, feudal, patrimonial, and other precapitalist economies would be transformed and incorporated into the global capitalist system'. Henry Paget, 'The Caribbean Plantation: Its Contemporary Significance', in *Sugar, Slavery, and Society: Perspectives on the Caribbean, India, the Mascarenes, and the United States*, ed. by Bernard Moitt (Gainesville, FL: University Press of Florida, 2004), pp. 157–87 (p. 158). The demographic of the indigenous population in Hispaniola remains uncertain. But, as Roger Plant demonstrates, the scale of the extermination is clear: 'By 1508, official Spanish census figures could count only 60,000 survivors. By the 1550s that number had been halved, and in 1570 contemporary documents gave a figure of approximately 500.' Roger Plant, *Sugar and Modern Slavery: A Tale of Two Countries* (London and Atlantic Highlands, NJ: Zed Books, 1987), p. 5.

14 J. Michael Dash, *The Other America: Caribbean Literature in a New World Context* (Charlottesville, VA. and London: University Press of Virginia, 1998), p. 27.

15   By the beginning of the nineteenth century, the Tropics, in particular, began to be seen in terms of a kind of a pagan sensuality, very different from Rousseau's vision of the divine in nature or the impulse to establish a social and moral order in the face of nature's disorder.

Dash, p. 31

16 According to J. Michael Dash, the view of nature as a route to sublime can be traced back to Jean-Jacques Rousseau. Later, similar images would resonate in Baudelaire's *Les Fleurs Du Mal* (1857), which presented nature as a place of worship. The portrayal of nature as potentially chaotic is manifested, meanwhile, in works like *The Tempest* and *Robinson Crusoe*. Dash, p. 29.

17 White goes so far as to claim that the concept of 'the Noble Savage' only served the ideological needs of the rising middle classes, who wanted to share in the 'natural' privileges that the nobility enjoyed and inherited. The claim to 'nobility' was only extended to the whole of humanity in principle as in fact it was 'meant to extend neither to the natives of the New World nor to the lowest classes of Europe, but only to the bourgeoisie' (p. 194). Hayden White, *Tropics of Discourse: Essays in Cultural Criticism* (Baltimore and London: Johns Hopkins University Press, 1978), p. 191.

18 Tzvetan Todorov, *The Conquest of America: the Question of the Other*, trans. by Richard Howard (New York, NY: Harper Perennial, 1992), p. 5, in Dash, p. 28.

19 Dash, p. 32.

20 Bryan Loughrey, *The Pastoral Mode: A Casebook* (London: Macmillan, 1984), p. 8.

21 Gifford, p. 4.

22 Gifford, p. 2.

23 Gifford, p. 2.

24 Theocritus's *First Idyll* does indeed contain motifs that are still now regularly regarded as inherent in the pastoral, such as 'idyllic landscape, landscape as a setting for song, an atmosphere of *otium*, a conscious attention to art and nature, herdsmen as singers, and, in the account of the gifts, herdsmen as herdsmen' (Alpers, p. 22). But as Renaissance reformulations testify, rather than simply reproducing a certain set of images and themes, pastoral as a genre has long been a historically diversified form. In fact, it is only in the Romantic period, following Friedrich Schiller's seminal *On Naïve and Sentimental Poetry* (1795–6), that the image of Golden Age nature, filled with a sense of innocence, nostalgia and loss, became the defining characteristic of the pastoral. As Paul Alpers explains:

> Nature and idyllic landscape figure prominently in most scholarly and critical accounts of pastoral and are regularly associated with the Golden Age, innocence, and nostalgia. It is not self-evident that these are the defining features of pastoral. The idea that they are derives from a specific poetics, which, so far as it concerns pastoral, has its profoundest statement in Friedrich Schiller's *On Naïve and Sentimental Poetry* (1795–6).
>
> Paul Alpers, *What Is A Pastoral?* (Chicago, IL. and London: University of Chicago Press, 1996), p. 28

25 Peter V. Marinelli gives the following example of the influence of the pastoral on other literary genres: '[Pastoral] appears not only in the specific form in which it came to birth (the "Grecian song" or idyll of Theocritus), but [also] interpenetrates the drama, as in Shakespeare's *As You Like It* and *The Winter's Tale*; the elegy, as in Spenser's *Astrophel*; and the epic, as in Spenser's *The Faerie Queene*'. Peter V. Marinelli, *Pastoral* (London: Methuen, 1971), p. 7.

26 Annabel Patterson, *Pastoral and Ideology: From Virgil to Valéry* (Berkeley, CA: University of California Press, 1987), pp. 8–10.

27 Gifford reiterates this double potential of the pastoral as both re-affirmation and critique:

> So the pastoral can be a mode of political critique of present society, or it can be a dramatic form of unresolved dialogue about the tensions in that society, or it can be a retreat from politics into an apparently aesthetic landscape that is devoid of conflict and tension.
>
> Gifford, p. 11

28 Graham Huggan and Helen Tiffin, *Postcolonial Ecocriticism: Literature, Animals, Environment* (London and New York, NY: Routledge, 2010), p. 85.

29 Lawrence Buell in his 1995 *The Environmental Imagination* explores the versality of the form in a distinctly New World context, which he terms the 'ideological pastoral'. For him,

> [t]his duality was built into Euro-American pastoral thinking from the start, for it was conceived as both a dream hostile to the standing order of civilization (decadent Europe, later hypercivilizing America) and at the same time a model for the civilization in the process of being built.
>
> Lawrence Buell, *The Environmental Imagination: Thoreau, Nature Writing, and the Formation of American Culture* (Cambridge, MA: Harvard University Press, 1995), p. 50

30 Marinelli, p. 11.

31 Marinelli, p. 12.

32 Gifford, p. 120.

33 Gifford, p. 152.

34 Huggan and Tiffin, p. 115.

35 Huggan and Tiffin, p. 115.

36 Huggan and Tiffin, p. 84.

37 Chris Park, 'environment', in *Oxford Dictionary of Environment and Conservation Online* (Oxford University Press, 2013) www.oxfordreference.com/view/10.1093/acref/9780199641666.001.0001/acref-9780199641666-e-2563 [accessed 9 March 2015].

38 Beverly Bell, *Walking on Fire: Haitian Women's Stories of Survival and Resistance* (Ithaca, NY: Cornell University Press, 2001), p. 23. Here, in the epigraph to the chapter, Bell quotes a remark by Yolette Etienne.

39 I would like to thank Sophonie Zidor for this reference.

40 Huggan and Tiffin, p. 117.

41 As Henry Paget makes clear, 'the plantation became the institution that provided the broad framework through which subsistence, feudal, patrimonial, and other precapitalist economies would be transformed and incorporated into the global capitalist system'. Henry Paget, 'The Caribbean Plantation: Its Contemporary Significance', in *Sugar, Slavery, and Society: Perspectives on the Caribbean, India, the Mascarenes, and the United States*, ed. by Bernard Moitt (Gainesville: University Press of Florida, 2004), pp. 157–87 (p. 158). The demographic of the indigenous population in Hispaniola remains uncertain. But, as Roger Plant demonstrates, the scale of the extermination is clear: 'By 1508, official Spanish census figures could count only 60,000 survivors. By the 1550s that number had been halved, and in 1570 contemporary documents gave a figure of approximately 500.' Roger Plant, *Sugar and Modern Slavery*, p. 5.

42 Marquez Stathis, p. 256.

43 See Beverly Bell, 'The Business of Disaster: Where's the Haiti-Bound Money Going?', *Other Worlds Are Possible*, 8 April 2010 www.otherworldsarepossible.org/business-disaster-wheres-haiti-bound-money-going [accessed 7 January 2014]. See also her book *Fault Lines* (2013), in particular Chapter 18, 'The Superbowl of Disasters: Profiting from Crisis' (pp. 146–53), gives a detailed analysis of the distribution of post-earthquake aid reconstruction contracts.

44 In his *Postcolonial Tourism: Literature, Culture, and Environment* (2011), Anthony Carrigan offers a highly original analysis, rooted in postcolonial studies and inter-disciplinary tourism studies, of the tensions of island tourism and how these are explored and staged in literary works. One of the key tensions is between employment and exploitation:

> Tourism propels environmental transformation, cultural commoditization, and sexual consumption – all processes that are acutely felt in many countries still grappling with the legacies of western colonialism. At the same time, tourism is consistently welcomed across the postcolonial world as a much-needed source of job creation and foreign exchange, even if the power relations that condition these transactions are distinctly asymmetrical.
>
> Anthony Carrigan, *Postcolonial Tourism: Literature, Culture, and Environment* (New York: Routledge, 2011), p. ix

45 After controversial elections, Martelly was elected president of Haiti in May 2011.

46 Haiti Grassroots Watch, 'Martelly Government Betting on Sweatshops: Haiti: "Open for Business"', *Haïti Liberté*, 29 November 2011 www.haiti-liberte.com/archives/volume5–20/Martelly%20government.asp [accessed 19 August 2013].

47 Contrasting analyses of such 'development' opportunities are best visible when comparing a report from Haiti Grassroots Watch (which overtly calls them a sourcing of slave labour) and an article from *The Economist*, which recognises Martelly's efforts: 'Mr Martelly may be right that to attract private investment Haiti needs to change its image of eternal aid supplicant into one of a hard-working place.' 'Open for Business', *The Economist*, 7 January 2012, www.economist.com/node/21542407 [accessed 19 August 2013]. See also Haiti Grassroots Watch, 'Haiti "Open for Business": Sourcing Slave Labor for US-Based Companies: Is the Caracol Industrial Park Worth the Risk?', *Haïti Liberté*, 13 March 2013 www.globalresearch.ca/haiti-open-for-business-sourcing-slave-labor-for-u-s-based-companies/5327292 [accessed 19 August 2013].

48 In addition, the real wages of the workers at the CIP, after the costs of transport and a midday meal, are 57 gourdes or approximately $1.36 per day. See Haiti Grassroots Watch, 'Haiti "Open for Business": Sourcing Slave Labor for US-Based Companies: Is the Caracol Industrial Park Worth the Risk?'

49 Phil Macnaghten and John Urry, *Contested Natures* (London: SAGE, 1998), p. 1.

50 Huggan and Tiffin, p. 117.

51 See the title of Potter's 2009 article: Amy E. Potter, 'Voodoo, Zombies, and Mermaids: US Newspaper Coverage of Haiti', *The Geographical Review*, 99 (2009), 208–30.

52 Homel, p. 82.

> Le séisme n'a pas détruit Port-au-Prince, car on ne pourra construire une nouvelle ville sans penser à l'ancienne. Le paysage humain compte. Et sa mémoire fera le lien entre l'ancien et le nouveau. On ne recommence rien. C'est impossible d'ailleurs. On continue. Il y a des choses qu'on ne pourra jamais éliminer d'un parcours: la sueur humaine. Que fait-on de ces deux siècles, et de tout ce qu'ils contiennent qui ont précédé l'année zéro. Les jette-t-on à la poubelle? Une culture qui ne tient compte que des vivants est en danger de mort.
>
> Laferrière, pp. 85–6

53 Derek Walcott, 'Nobel Lecture: The Antilles: Fragments of Epic Memory', *Nobelprize.org* (Nobel Media AB: 2014) www.nobelprize.org/nobel_prizes/literature/laureates/1992/walcott-lecture.html [accessed 13 November 2017].

54 Walcott, 'Nobel Lecture'.

55 Homel, p. 118.

> Cette situation a créé une nouvelle réalité à laquelle il fallait rapidement s'adapter. «Tu vois où se trouvait le Caribbean Market? Alors tu continues un peu jusqu'à

> passer deux immeubles par terre …» A la réalité de cette ville en miettes, les gens
> ont ajouté des éléments de l'ancienne ville qui flotte encore dans leur mémoire.
>
> Laferrière, p. 124

56 Walcott, 'Nobel Lecture'.
57 Homel, p. 122.

> Pourquoi ne pas penser à peindre certains quartiers? A faire de Port-au-Prince une
> ville d'art où la musique pourrait jouer un rôle? Haïti doit profiter de cette trêve
> pour changer son image. On n'aura pas une pareille chance (façon de parler) une
> deuxième fois. Présenter un visage moins crispé.
>
> Laferrière, p. 129

58 For an overview of the project and the initial reactions to it, see Marie-Julie Gagnon,
    '«Jalousie en couleur»: transformation extrême d'un bidonville', *Le Huffington Post*
    Québec, 16 May 2013 http://quebec.huffingtonpost.ca/2013/05/16/jalousie-haiti_n_
    3287761.html [accessed 19 August 2013]; 'Haïti – Reconstruction: Deuxième phase
    des travaux de rénovation du quartier de Jalousie', Haiti Libre, 15 August 2013 www.
    haitilibre.com/article-9228-haiti-reconstruction-deuxieme-phase-des-travaux-de-reno-
    vation-du-quartier-de-jalousie.html [accessed 20 September 2013].
59 The term 'poverty tourism' or 'poorism' has been introduced in order to conceptualise
    'a new form of tourism that has emerged [since the mid-1990s] in the globalising
    cities of several so-called *developing countries* or *emerging nations*'. One of the
    essential features of this form of tourism is visiting the most disadvantaged parts of a
    respective city. Among the most popular destinations are townships in Cape Town,
    favelas in Rio de Janeiro and the Dharavi informal neighbourhood in Mumbai. Other
    terms include 'slumming, 'reality tours', and 'cultural and ethnic tourism'. These, to
    varying extents, can be seen as linked to the wider category of 'dark tourism': see
    Manfred Rofles, 'Poverty tourism: theoretical reflections and empirical findings
    regarding an extraordinary form of tourism', *GeoJournal*, 75 (2010), 421–42 (p. 421).
    One of the key paradoxes of poverty tourism or slum tourism (also referred to as
    'slumming') is the claim that this type of activity can contribute to poverty alleviation
    in a given neighbourhood. But '[i]f slum tourism was a successful strategy for poverty
    mitigation, would it not undermine its own premise?' See Fabian Frenzel, 'Slum
    tourism in the context of the tourism and poverty (relief) debate', *Die Erde: Journal
    of the Geographical Society of Berlin*, 144 (2013), 117–28 (p. 118). See also Fabian
    Frenzel and others, 'Development and globalization of a new trend in tourism', in
    Slum Tourism: Poverty, Power and Ethics, ed. by Fabian Frenzel, Malte Steinbrink,
    and Ko Koens (New York, NY: Routledge, 2012), pp. 1–18.
60 There have been other, less controversial projects carried out in Jalousie which mark-
    edly improved the welfare of the local population, such as the Projet Jalousie part of
    UNESCO's Projet Villes/ MOST Cities Project (1996–2001) which was implemented
    in Dakar, Senegal and Jalousie, Haiti. In Jalousie, inhabitants of the neighbourhood,
    under the guidance of Patrick Vilaire, a Haitian sculptor and painter, created a
    30-metre-long mural that leads to a source of water where they need to pay only one
    gourde per bucket. Before, young girls had to fetch water far away and were often
    victims of rape. This was a multi-actor project with a number of local organisations
    and NGOs, with UNESCO acting as a 'catalyst' to legitimise the initiative. Together
    with the creation of the football field, public squares lit by night and a pedestrian
    bridge, the mural significantly contributed to the improvement of the local infrastruc-
    ture and social interactions. For a detailed description and evaluation of the project,
    see Denis Merklen, *Management of Social Transformations MOST: Urban Develop-
    ment Projects: Neighbourhood, State and NGOs, Final Evaluation of the MOST Cities
    Project* (UNESCO, October 2000) www.unesco.org/most/dp54merklen.pdf [accessed
    29 December 2014].

61 'Favela Painting Foundation', *Favelapainting.com*, www.favelapainting.com [accessed 16 November 2016].

62 'Favela Painting Investigation in Haiti', *Akvo.org* https://rsr.akvo.org/en/project/1855/#report [accessed 20 December 2016].

63 Sibylle Fischer, 'Fantasies of Bare Life', p. 2. Increasingly, however, it seems that the project is merely sugar-coating the neighbourhood's deep-rooted problems such as lack of housing and the constant threat of demolition and displacement, some of it a direct cause of the June 2012 protests. For an analysis of the historical construction of Port-au-Prince's vulnerability, amplified by a lack of housing codes, intense migration to the city under the Duvalier regime, and destructive agricultural policies, see Kathleen A. Tobin, 'Population Density and Housing in Port-au-Prince: Historical Construction of Vulnerability', *Journal of Urban History*, 39 (2013), 1045–61. See also Al Jazeera, 'Haitians Protest Home Demolition Plans', www.aljazeera.com/news/americas/2012/06/201262665643926283.html [accessed 09 September 2013].

64 Homel, pp. 157–8.

> Chacun a le droit de savoir dans quel genre de ville il aimerait vivre. Et mieux, il devrait pouvoir intervenir dans l'élaboration du plan de la nouvelle ville. […] Le matériau le plus important, c'est encore l'esprit. Un esprit qu'on voudrait voir tourné vers le monde, et non replié sur lui-même. […] Une nouvelle ville qui nous forcerait à entrer dans une nouvelle vie. C'est cela qui prendra du temps. Ce temps qu'on refuse de s'accorder.
>
> Laferrière, pp. 166–7. Homel, pp. 157–8

65 'palimpsest, n. and adj.', *OED Online*, Oxford University Press (2018) www.oed.com/view/Entry/136319 [accessed 22 February 2018].

66 This point is made in Alfred W. Crosby's seminal work: *The Columbian Exchange: The Biological and Cultural Consequences of 1492*. In it, Crosby explores the impact that exchanges of food and disease between the Old and New Worlds had on respective ecosystems and lifestyles. See Alfred W. Crosby, *The Columbian Exchange: Biological and Cultural Consequences of 1492* (Westport, CT: Greenwood Press, 1972).

Part IV

Disaster selves

8   Narrating conversion and rescue

The tremors in Haiti may have ended, but the experience of catastrophe is far from over. Like the permanently changed landscape, the survivors' sense of self has been left unsettled, with the earthquake redefining the significance of all previous life-events. Yet prevailing images of wholesale destruction and homogenising portrayals of those affected by the disaster risk masking natural histories of the landscape and the pre-existing social inequalities that significantly shaped responses to the event – the tremors may have flattened buildings, but they did not level social inequalities and dynamics of power. The experience of the event's impact was anything but universal. The only affinity between a foreign humanitarian worker who had just arrived in Haiti and an inhabitant of one of Port-au-Prince's informal neighbourhoods was the sense of being *there* and *then* at the same time, as a direct witness to the earthquake; in Laura Rose Wagner's gloss, 'fault lines [were] social as well as geological'.[1] In such a context, what does it mean to narrate one's own experience of the disaster or to create a fictional narrative out of it? How are one's subjectivity and sense of collective agency redefined in the process? What possibilities are offered for the Haitian collective? What collective futures are imagined? What are the ethical and political dimensions of narratives as they attempt to offer post-earthquake definitions of individual rescue and community recovery and healing?

These fundamental questions are central to Dan Woolley's autobiographical conversion account *Unshaken: Rising From the Ruins of Hotel Montana* (2011) (thereafter *Unshaken*), Nick Lake's *In Darkness* (2012), a fictional narrative of a boy trapped under hospital rubble, and Laura Rose Wagner's *Hold Tight, Don't Let Go* (thereafter *Hold Tight*) (2015), a story of two cousins' life after the earthquake. An analysis of these texts' narrative aesthetics and divergent meaning-making frames is an entry point into an examination of the unsettling effect the disaster had on categories of self and subjectivity – foundational components of individual experience and narrative form – and ideas of individual rescue, collective and diasporic recovery and healing possible in the aftermath of the disaster. The three aesthetically and formally diverse narratives explore transformations triggered by the earthquake within the context of an individual's life and the wider histories behind the 2010 tremors. In their respective engagement with evangelical Christian theology, Vodou cosmology and symbolism, defining events in the Haitian history, such as the 1791–1804 Haitian Revolution, as well as contemporary Haitian culture and everyday experiences, the three texts use twinning, as a metaphor and an organising principle, in an attempt to conceptualise the experience of the 2010 disaster and its place and significance within the intertwined, personal and collective, chronologies. In what follows, the chapter first looks to Woolley's autobiographical text of spiritual rescue and renewal to then explore two fictional accounts of post-earthquake twinned lives as imagined in Lake's and Wagner's young adult novels. While focusing on the impact of the disaster on individuals, these texts equally inspire a consideration of ideas of community fragmentation and collective recovery and healing in the aftermath of the earthquake: 'how sudden moments of rupture and crisis cause people and communities to come together or to come apart'.[2] By embedding these issues within an interrogation of the country's history, the many breaks and re-assemblages the earthquake forces, Lake's novel most specifically seems to ask, in Colin (Jean) Dayan's gloss: 'Where can one go to find the proper history that can help to correct the present? How can one write in a belittered world?'[3] Whether envisaging the earthquake as a direct cause for subjective transformation and renewal, understood for Woolley through a specifically religious framework, a general metamorphosis and the ontological process of becoming someone else, as envisaged by Lake, or a forced acceptance of absence and separation, as explored by Wagner, the three narratives offer a complex reworking of recovery not as a return to a pre-disaster past or a teleological progress, but rather as an open-ended, iterative and itinerant process of slow healing.

Twinned crises, double rescue

Questions of self and subjectivity are first introduced in Dan Woolley's *Unshaken* through its structure and paratextual features such as preface, epigraphs, chapter sequence and epilogue. These embody the aims of the text and complement its thematic emphasis on coming to terms with the event and envisaging new post-earthquake definitions of subjectivity. *Unshaken* is a highly autobiographical account, focused, at least initially, around the story of Dan Woolley who came to Haiti just a couple of days before the earthquake struck on a short assignment to produce video material for Compassion International – a large international evangelical charity. Compassion International is a Christian child advocacy ministry that 'releases children from spiritual, economic, social and physical poverty and enables them to become responsible, fulfilled Christian adults'.[4] The charity was founded in 1952 in South Korea by American evangelist Rev. Everett Swanson, who 'felt compelled to help 35 children orphaned by the Korean conflict',[5] and now has over 6,214 international church partners and sponsors over 1.8 million children worldwide.[6] Woolley was in Haiti to make a video showcasing one of Compassion's projects in Haiti, the Child Survival Programme, in order to encourage current and potential donors to support the organisation's work. This specific context for Woolley's presence in Port-au-Prince, one linked to the continued expansion of Protestant missions in Haiti,[7] directly shapes the text's aesthetics as well as its consideration of subjective recovery and community healing.

From the outset, the autobiographical narrator is very clear about his work and the task he needs to accomplish, namely to create a video clip for Compassion International 'that would move donors to care about mothers and babies they'd never met in places they'd never been'.[8] On the morning of 12 January 2010, Woolley joins a local pastor from a partner church in order to film a group of mothers and children who have gathered in the church hall. His search for 'the right story' is marked by a sense of urgency; for if it is to be successful in attracting contributions, the video must create an intimate relationship between the mothers and children from the congregation as well as potential donors in the U.S. and elsewhere:

> Most donors will never get within three hundred miles of the poverty in Haiti, but if they can watch a video on their computer that gets them even three steps closer to a mom who lives it every day, then we will have done our job well.[9]

During the day, Woolley spends his time filming women at a local church centre. Just after entering Hotel Montana, an upmarket hotel located on the hills of the affluent Pétionville neighbourhood where he is staying during his trip, the earthquake strikes and the hotel collapses, imprisoning Woolley and fellow members of his team. While reflecting on his chances of being found and rescued in time, the narrator does briefly acknowledge the socio-economic tensions that the Montana (Part III), known for its high-class international clientele (Bill Clinton and Brad Pitt, among others), embodies. He also engages, albeit briefly, with the thorny question of economic disparities between foreign NGOs and the local population they serve:[10]

> I wasn't ethnocentric enough to think I deserved any special treatment because I was a white American. In fact, I thought quite the opposite. The Haitians would already have so much on their hands that they should pay attention to their own people first. I was a guest in their country, staying where few Haitians could afford to stay. If I was going to be rescued, it was more likely that Americans would rescue me.[11]

With this self-reflexive remark, the narrator introduces a level of contextual consideration into his narrative, otherwise omitted in the highly focalised spiritual account, putting to the fore the entanglement of aid, privilege and power as reflected in unequal mobilities and access to space. Paradoxically, these are the very dynamics and exclusions that, in the long term, Woolley's charitable work hopes to level. The recognition of one's privileged position within these very structures is, however, only a momentary one. All previous aspirations set aside, the narrator's main objective now is to survive until the arrival of the rescue teams, regardless of where they are from. The notes he makes in his notebook during this time of being imprisoned in the ruins of the hotel provide the substance of his later published account.

In *Unshaken*, Woolley uses a religious frame for the narrative of the few hours preceding the disaster, its immediate aftermath and the time he is trapped in the Montana, adding to his personal account of surviving the earthquake a retrospective reflection on his marriage and his wife's depression, from which she suffered over a six-year period from 1992 to 1998. Radically different, both crises are seen together as times of trial that can only be comprehended within a religious frame and a personal trajectory of spiritual conversion and growth. For Woolley then, as the preface already makes evident, the two events are key milestones in his spiritual journey, equally functioning as tangible proofs of God's presence in the narrator's life, an experience he is compelled to share:

> Some may wonder why I have chosen to share this story of survival and rescue, especially since the stories of many others impacted by the earthquake did not end well, at least in human terms. [...] I tell my story because I had an encounter with God in the midst of this crisis [...] I feel compelled to give voice to my experience and testify to the grace of the God who was with me in the depths of my ruin.
>
> It is for this purpose – to glorify God – that I was created, and this is the only reason for this book.[12]

In this extended quotation, the aims of the text, composed of the narrator's memories of his marital difficulties and his survivor account, are expressed in strong religious terms. In turn, the narrative itself becomes an act of praise that glorifies '[the] unshakable God'[13] with the authorship of the book being partly divine: the author-narrator shifts the focus of this conversion account from the two episodes of his own suffering, and the events themselves, towards their wider theological significance and resonance. In this context, the adjective 'unshakable', from the book's cover, refers, contrary to initial impressions, not so much to Woolley's faith as to God's presence; for the author's religious beliefs are indeed shaken and put to the test on numerous occasions during the course of the account. The narrative is thus at once a story of religious doubt (and its eventual overcoming) and a form of address to the Creator, who constitutes the only certainty in the author's life. Woolley's narrative duly acquires a psalmic character: it is as much a testimony of physical as of religious survival, of rising from the material and spiritual ruins towards a new, potentially better, life.

In his attempt to share and make these profound experiences accessible and to locate their meaning, *Unshaken* draws on the easily identifiable, yet far from homogenous, Christian tradition of conversion narratives which can be traced back to St Augustine's *Confessions* (*c.*398–400).[14] Openly religious, Woolley's text embraces the guiding aim that defines many evangelical accounts, namely 'the intent itself to convert, to propagate the kingdom of the faithful by persuading unconverted readers and listeners of their own need for redemption'.[15] Yet, apart from this broad characteristic, conversion accounts are defined by diversity of form and by complex variations in the theology and spiritual traditions they endorse. In the context of Woolley's account, for example, it is important to note that his is a work of spiritual renewal and strengthening of religious zeal rather than the rendering of an original conversion experienced by a non-believer. In effect, his is an account of a re-encounter with God which, still, draws on the rhetorical conventions of evangelical narratives, framing two traumatic experiences (his troubled marriage and the 2010 earthquake, respectively) as parallel moments of crisis and rescue that lead to being 'born again'.

In this dual context, *Unshaken* hopes to return to 'the truth' of faith while crafting a carefully designed yet genuine narrative of the recreation of self for the reader, one that presents 'an integrated, continuous personality which transcends the limitations and irregularities of time and space and unites all of his or her apparently contradictory experiences into an identifiable whole'.[16] To this end, the text brings together two distinct experiences, Woolley's wife's depression and his entrapment under the rubble, via a double metaphor, namely the transition from darkness to light and from fragmentation to wholeness. In effect, *Unshaken*, at once 'intimately autobiographical'[17] and committed to a system of beliefs,[18] doubles as an autobiography and a conversion account in the evangelical tradition and is simultaneously a work of narrative *reconstruction* and one of subjective *recovery*: it is both a '[r]etrospective prose narrative that a real person makes of his own existence'[19] and 'an artistic arrangement of facts, an imaginative organization of experience with an aesthetic, intellectual and moral aim'.[20] By pairing up these two distinct moments in his life and re-interpreting them in the light of his strengthened religious outlook, the narrator affirms that conversion, as well as attempts at meaning-making, is 'an iterative process': one that might well begin with a strong and identifiable moment of epiphany, but can only be completed through the believer's sustained commitment to recovered values.[21]

Preceding and foreshadowing the main narrative of conversion and rescue, as found in the text, the chapter sequence – with some chapters authored by Dan's wife, Christy – already suggests this double trajectory, emphasising the importance of the first crisis for Woolley's experience of conversion under the rubble. It is the narrator's awareness of his wife's vulnerability, following years of her struggle with severe depression, and his anxiety about his family, made explicit in the book's opening dedication, that originally motivated him to write down his thoughts, experiences and words of advice while he was waiting to be rescued from the hotel debris. The first words the author-narrator jots down in his diary are words of farewell and spiritual counsel to Christy and their two sons, Josh and Nathan, with photos of these first pages written in the dark included in the printed edition of the book. Soon after, he turns to everyday considerations and writes 'lists of practical things, like how to access [their] online banking information, passwords for [his] e-mail accounts, and details of how to access [their] assets and pay bills'.[22] As he lies buried under the ruins of the hotel, Woolley reflects on his wife's depression, which started soon after they got married, and this renewed realisation of Christy's emotional fragility as well as the financial difficulties she would be faced with following his death inspires him to seek other ways of being rescued. Rather than passively waiting for the arrival of the relief teams, he decides to leave the lift-shaft in which he is hiding and explore the surrounding rubble in the hope of finding an alternative exit and establishing contact with other survivors. In this retrospective reading of events it becomes clear that, paradoxically, this first longer crisis becomes a life-saving one, engendering Woolley's resolve to seek out the rescue troops.

In addition to the pairing of these two threads, which complement and mirror each other throughout the account, the book's epilogue ties the two in a surprising comparison that is followed by words of direct advice to the reader. Revisiting the two disasters in his life, the narrator comes to realise that it was indeed this 'other rescue'[23] of his wife from depression that was the more significant with the relative brevity, approximately 65 hours, of being buried under the rubble contrasted with the six long years of his wife's depression. Here the difference in duration seems to correspond to the weight of the emotional impact these former events had on Woolley and his wife. This realisation is quickly followed by a call to the reader who, the author assumes, might also suffer from depression or know someone who does: 'If you or someone you know is experiencing symptoms of depression, please seek professional help. Rescue from *your* darkness is possible too' [italics added].[24] While reaffirming the narrative design of the account, these paratextual features contribute to the sense of intimacy and mutual trust between the narrator and the reader. With this intimate appeal, anchored in his own experience, the narrator also hopes to trigger direct action on the part of the reader, e.g. seeking professional help or encouraging others to do so. This gesture of reassurance, similar to his notes of advice to Christy, reaffirms the metaphorical movement from darkness to light in Woolley's narrative, extending it far beyond the site and time of the spiritual epiphany.

However, the conventions and the personal focus the narrative adopts do not necessarily preclude a wider consideration of the catastrophe and its many afterlives. *Unshaken*'s aim to be an inspiration for others and a trigger for change is intricately linked to its fundraising appeal – the narrative's way of complementing and translating its earlier call to individual conversion and of facilitating long-term material community recovery through the reader's response. In the paragraph following upon the comparison of the two 'dark moments' in the narrator's life, the author openly endorses the charity's public work: 'Finally, I encourage you to support Compassion International [...]. Learn more about [...] how *you* can make a difference in the lives of mothers, babies, and children trapped in poverty' [italics added].[25] The accompanying address to the reader, which follows the author's acknowledgements, provides a comprehensive outline of the range of activities that fall under Compassion's Child Survival Programme. In an accumulation of staccato sentences, Woolley underlines the impact each donation can make, extending the metaphorical trajectory of the narrative, from darkness to light, and the titular image of rising from the ruins: 'Your support can literally mean the difference between life and death for a young vulnerable child. [...] If you are moved by the needs of children in poverty, please do something about it. Today. Right now.'[26] The tone of the address is overwhelmingly emotional, aimed at increasing the reader's sympathy for the plight of Haitian women and children while inspiring his/her active response. It is no longer only the narrator or the reader who needs to be rescued from darkness, but, thanks to an individual donation, entire communities can be given a chance to escape the all-engulfing trap of poverty, disease and malnutrition.

As such, the 'success' of the narrative extends beyond its particular aesthetic or spiritual goals, can be measured in tangible ways – namely donations made to Compassion International – and relies on maintaining a balance between distance and proximity: Haiti may be too far away to concern us, but now the scale of the suffering makes it 'close to our hearts'. In this way, the text seeks to mobilise a more emotionally invested response from the target audience who can distinguish and identify with familial themes, shared doubts or with recognisable 'Third World' images of poverty, innocence and vulnerability, even if the complexities of the 2010 disaster might escape, or possibly even not interest, them. This carefully crafted emotional proximity transforms the significance of donating; it is no longer just a gesture of financial support but instead becomes a means of making a real change and, for those who share or are inspired by the book's spiritual message, a direct expression of the Gospel call to charity. Consequently, by answering the book's plea and helping the work of Compassion International, the reader is co-authoring new stories of rescue and salvation. In this context, her/his donation has equally a cathartic dimension: by contributing to the relief effort she/he experiences personal relief at having responded to the call for help, reaching out to the affected other.

The aspirations and guiding aims expressed by the text's formal and extratextual features directly shape and are twinned with the account's thematic exploration of individual and collective transformation and recovery in the wake of the January 2010 disaster. For the narrator of *Unshaken*, being saved from the rubble is a manifestation of God's omnipotence and an unrepeatable opportunity, which Woolley duly seizes, granted to him by God to live a full life of total devotion. The protagonist's individual metamorphosis and spiritual renewal triggered by the earthquake have their beginnings in the experience of the liminal state of survival. Trapped under the debris of the Hotel Montana, he battles with feelings of extreme uncertainty and incomprehension. First, he tries to understand what has happened, then he attempts to come to terms with the fact of his random survival. Yet as time passes, the chances of being rescued steadily decrease and this realisation forces him to confront the increasing likelihood of death. It is at this moment of profound unease that spirituality, here framed in the evangelical Christian tradition, helps him to negotiate the liminal position he occupies, to find meaning in his suffering and to intuit a parallel between this recent ordeal and earlier times of crisis. The text's narrative arc reaffirms the coupling of these two processes of subjective reconstruction and clearly demarcates the main stages of Woolley's spiritual journey: first comes the experience of profound spiritual doubt and despair; second is the act of turning to God, a moment of personal conversion followed by a gradual acceptance of whatever outcome lies ahead. At heart, this is a narrative of progress and spiritual advancement and a movement from fragmentation to wholeness. Still, the iterative and open-ended nature of this metanoia undermines the sense of completeness or finality of this spiritual journey.

An all-encompassing (both metaphorical and literal) darkness marks the first stage of this double struggle for survival and spiritual renewal:

> I spit out the blood and dust that coats my mouth but I can't spit out the fear. […] I'm hanging on to the realization that I lived through an earthquake. I survived! But I also know that if I want to make it out of this black tomb alive, if I ever hope to see my family again, it will take a miracle – a series of miracles.
>
> Miracles I'm not sure I have the faith to believe in.[27]

Here, the sense of anguish and utter confusion are mixed with the narrator's self-assurances of hope and faith. The ambiguous tone of the passage reveals his uncertainty over whether to rejoice or not at having survived the earthquake: although he is still alive, he doubts the possibility of ever being rescued from the rubble and, in effect, is forced to contemplate the ever-increasing likelihood of his own death. At the same time, the two images have a clear religious resonance as they evoke Christ's resurrection and the biblical stories of Lazarus,[28] the widow's son[29] or Jairus's daughter[30] being raised from the dead. This pairing contrasts with the engulfing darkness of the tomb, prefiguring the narrator's possible death with the imagery of light and vision, suggesting the chance to see his family once again. He hopes that he too can be rescued from darkness, brought back to life by God, and be reunited with those he loves. Yet as the two closing lines of the passage make clear, this hope is a fading one; the narrator is no longer certain that divine intervention can occur. At this stage, his faith fully depends on one particular 'miracle', namely whether he can be rescued from the rubble or not.

However, as the initial experience of shock gives way to a contemplation of his current predicament, the narrator begins to revise his initial response to the aftermath of the earthquake. There is a constant tension at work here. On the one hand, the likelihood of his survival decreases as the hours since the disaster pass; but on the other, he refuses to fully give up hope in his rescue and continues to struggle to find meaning for the testing events in which he is embroiled: 'I had lost a lot of blood; there was no food or water; and the aftershocks continued. I tried to postpone my death in every way I could, but I knew very little was up to me.'[31] Left with little, if any, possibility to impact his situation, the narrator totally depends on the external help of the rescue teams who, God willing, will save him and those buried with him in Hotel Montana. It is this sense of powerlessness and dependency that triggers the narrator's re-evaluation of his life. Proximity to death makes him re-assess the strength of his commitment to faith, which is the only thing, however uncertain, to which he can now turn. Recalling the words of the Book of Revelation,[32] Woolley realises that his heart 'was luke-warm – neither hot nor cold toward God'[33] and that, as a result, he had previously led a 'pretty standard, mediocre Christian life'.[34] With the biblical call to repentance echoing in his mind, Woolley is quick to admit that the comparable sense of security he had enjoyed before coming to Haiti has resulted in his spiritual lukewarmness, making him more distant from God: 'In the face of death and eternity, [he] could not lie to [himself]. Something in [his] soul, in the core of [his] being, was off-kilter, and [he] knew it.'[35]

Echoing the increasing emphasis within the evangelical tradition on the ongoing character of metanoia,[36] Woolley's narrative suggests that the original experience of spiritual awakening is indeed repeated, throughout life, as a 'cyclic pattern of conversion and reconversion as if the converted [is] predisposed to repeat and reinforce this fundamental experience over and over'.[37] Within this pattern, any new challenges and doubts are recast retrospectively as ultimately enriching trials where subsequent moments of rerouting life's unexpected if painful turns which lead to a turning back and re-rooting in the divine. Consequently, the composing of a conversion narrative, which paradoxically aims at a sense of finality and cathartic closure, is constitutive of the back-and-forth process of conversion itself. Moreover, by sharing his own story of deliverance and survival, the narrator hopes that the reader, moved by his testimony, will too restore his/her sense of purpose in life and be inspired to help others rebuild their lives. This moment of awakening, which follows upon Woolley's re-examination of his lukewarm life before his arrival in Haiti, transforms the narrator's initial reading of the situation: what needs to change in order for him to be truly 'saved' is his interior predisposition, not necessarily the material circumstances in which he finds himself.

Faced with the sobering realisation of his separation from God, the narrator desperately turns to prayer in the hope that this can alleviate his pain, help him realign himself with the will of God, at least for the last hours of his life, and find meaning in his current suffering. He is yearning for a sense of comfort, however limited, and indeed he soon experiences the moment of reassurance he is longing for: 'And while I worshipped and prayed, I heard a voice in my head say: *You are mine!*' [italics original].[38] This moment of epiphany and experience of a clear divine call starkly contrasts with the lack of communication with the search teams and the sense of Babel-like chaos[39] that characterises the early rescue efforts; sounds of 'saw from above', 'voices to [the] left' in French, English and other languages unrecognisable to the narrator create an overwhelming cacophony.[40] Powerless and desperate, Woolley's only hope comes from his belief in God's promise. Surrender and devotion define the narrator's renewed understanding of self and subjectivity: only by emptying himself of his individual desires and aspirations can he grow closer to God. The initial, involuntary experience of powerlessness in the face of disaster is now replaced, thanks to prayer, by conscious abandonment to the will of God.

However, this seemingly gradual progression towards a greater sense of spiritual trust is repeatedly interrupted by an equally intense experience of doubt, one which is never fully dealt with in the account. At the source of this mistrust is the recurring question of the purpose of his and his family's suffering and the narrator's profound hesitation whether 'any of this be good'.[41] Although this and other remarks in the text point to a longstanding Christian philosophical-theological debate on the meaning and sense of suffering and theodicy, the narrative does not engage explicitly with these scholarly arguments as they might alienate the book's potential target audience. In effect, they might too distract the reader from the main purpose of the account, namely to create a sense of trust and proximity between the narrator and the reader and to act as a personal testimony of faith and a confirmation that 'life-changing encounters with God can, do, and should take place'.[42] Emphasis is therefore placed on the deeply personal character of this spiritual trajectory and those aspects of faith that are common across evangelical denominations such as reliance on scripture, as manifested in the text through frequent recourse to biblical passages, and the concern for conversion, as expressed in the narrative arc of the account.[43] Paradoxically though, this repeated emphasis on individual struggle and personal experience of spiritual and physical deliverance – *Unshaken*'s dominant narrative framing – equally raises many unanswered questions about the narrator's own ability to lead a balanced life in the wake of the disaster and the significance of the event for Haiti and its people.

Rather than being fully resolved, the tension between faith, doubt and purpose is smoothed over through the use of analogy and metaphor as the account progresses and draws to a close. The narrator once again reaffirms the parallel between his experience of the earthquake and his wife's depression and goes so far as to conclude, as he did in the epilogue, that Christy's illness was the more traumatic of the two events. Only after this second experience of the 2010 disaster can the narrator effectively reassemble and reorder the distinct episodes and connect them into a recognisable and reassuring design. In this rereading, these testing experiences are given a clear sense of purpose, allowing Woolley to grow in faith and trust in God. Both the six years of his wife's depression and the 65 hours of being trapped in the Montana become life-lessons which have brought him closer to the fullness of life: 'With God's grace, I am no longer living a half life; instead, I am living a new kind of life.'[44] This double (structural and formal) parallel, which allows him to focus almost exclusively on the personal dimensions of these crises, suggests that his current ordeal will ultimately lead to something positive.

Even the permanent wounds that the earthquake leaves are seen mainly in physical terms, are given a positive significance and seem to represent the triumphant spiritual combat against all-engulfing darkness, with the narrator emphasising his recovery of faith and successful 'readjustment' to reality: 'The battle scars on my leg and the back of my head [...] serve as tangible reminders of the lessons I've learned, and I am grateful for both.'[45] Etched onto the narrator's body, these scars become ornamental reminders of two victorious battles on the part of a 'good and faithful servant'.[46] In metaphorical terms, the first stage of his imprisonment is a time of darkness and deafening cacophony; the second is the experience of light and a clear call from God that come from the joy of being reaffirmed in one's beliefs and, much later, the hope of seeing his family again. Foundational to this twinned trajectory, the two life-events function effectively as two complementary narratives of conversion and renewal.

Converting the collective

Moreover, the text's reading of the January events as a traumatic yet ultimately enriching experience that triggered spiritual improvement builds on the conviction of mandatory repentance for a previously misguided life – a call that might well be answered on an individual level but is harder, if not impossible, to realise by the collective. This interpretation evokes the 'deuteronomistic view of history embraced in many literal readings of the Bible',[47] which see disasters and curses as punitive judgement against those who do not obey God and do not adhere to His commandments.[48] Such a way of responding to disasters stresses the spiritual causes of material events, for example the earthquake. Within this causal theology, as promoted for example by the Christian New Apostolic Reformation and other evangelical congregations, the disaster and its aftermath figured as 'God's punishment of a sinful nation whose church is divided, whose people worship demons, and whose government is corrupt'.[49] The emphasis, both here and in Woolley's text, is placed on spiritual rather than material determinants of events, providing a meaningful interpretative frame for the disaster, one that places the event within an identifiable cosmology and discernible and traceable salvation history.

Although such causal framework can indeed provide a sense of meaning and purpose of the overwhelming event for an individual, it risks overwriting the particularities of the 2010 disaster, contributing to the dehistoricisation of the earthquake with only minimal consideration given to the long-term causes of post-earthquake damage and its lasting impact on individual and collective lives. In this context, the narrative's almost exclusive focus on testimony and the call to conversion, consistent with the generic conventions it embraces, is a key limitation to the text's vision for collective post-earthquake renewal. Instead, Haiti's material impoverishment is easily coupled with spiritual crisis. Only a blanket conversion to a very specific type of evangelical Christianity – regardless of the fact that the majority of Haitians already declare some type of religious affiliation[50] – can save the country from future disasters; an image epitomised in the scene of a fellow Haitian survivor's, Lukeson's, conversion under the rubble and his 'accept[ance] [of] Jee-sus' [*sic*][51] into his life. This dramatic moment of his friend's metanoia is neatly slotted into the overwhelmingly positive pedagogical frame of *Unshaken* with little, if any, acknowledgement of the extreme psychological vulnerability of his friend, and other victims of disaster, at the desperate moment leading up to one's religious experience. Rather, for both Woolley and Lukeson, the cataclysm becomes – quite literally – a blessing in disguise. With renewed faith and a new set of resolutions, both emerge unscathed and enriched from the Sheol,[52] ready to start a new chapter in their lives.

The scale of the 2010 earthquake indeed had a great impact on Haiti's religious landscape yet these new forms and expressions of religious belonging, as Karen Richman stresses, were characterised by fluidity of denominational affiliations and conversions, and were often determined by pragmatic rather than by theological motives: '[s]trategic positioning for purposes of personal advancement and/or protection, rather than deep conviction in the superiority of Protestant doctrine, is the reason many convert to Protestantism'.[53] In no way is this affiliation, whether to a Protestant group or to the Catholic Church, a fixed and stable one. Nor does it necessarily put an end to one's belief in the power of traditional Vodou practices or charms.[54] Fully aware of these socio-economic and psychological factors,[55] local Haitian Protestant pastors speak of a wave of 'bad conversions' – ephemeral and motivated by fear and panic rather than a lasting commitment to faith – that the earthquake triggered.[56] In Woolley's narrative of his spiritual trajectory, the singular emphasis on the experience of providence mutes such entanglement of privilege, depravation and grace, with the moment of enlightening epiphany, at least initially, overshadowing the current trials and any future difficulties or structural constraints to this 'full' life.

Consequently, *Unshaken* prioritises individual conversion (and/or donation) over the attempt to provide a sense of proximity to other survivors, as experienced, for example, by Martin (Part II). Whereas Martin unites his suffering with that of other victims affected by the disaster, through the prayer form of his account, Woolley seeks to confine the trauma of the event by emphasising the realignment of one's life. In effect, Woolley's account of his double survival, from rubble as much as from non-belief, gives an expression to a highly individualised interpretation of the January earthquake. This explanatory logic locates meaning and wider significance of the 2010 disaster, as well as its collective resonance, within a layered frame, one that, seemingly, seamlessly ties together personal life, divine designs and history of salvation, as made manifest in the here and now of the 2010 disaster. In so doing, the twinned form and thematic content of the account aim to join the two moments of crisis, connect the reader to Haiti, and to the divine through a teleological and theological account of conversion and change. Still, despite the narrator's affirmations, this knot is not a firm one. The certitude and joy of individual rescue are tainted by the awareness of the lack of a clear way out, towards a full life, for the collective, left stranded in the vulnerable everyday.

Notes

1 Laura Wagner, 'Haiti: A survivor's story', *Salon*, 2 February 2010, www.salon. com/2010/02/02/haiti_trapped_under_the_rubble/ [accessed 1 February 2017].
2 Laura Wagner, e-mails with the author, 3 February 2017.
3 Colin (Jean) Dayan, 'The Gods in the Trunk, or Writing in a Belittered World', *Yale French Studies*, 128 (2015), 92–112 (p. 112).
4 Compassion International, *About Us* www.compassion.com/about/about-us.htm [accessed 22 July 2015].
5 Compassion International, *History* www.compassion.com/history.htm [accessed 22 July 2015].

6 Compassion International, *Compassion International: Releasing Children from Poverty in Jesus' Name* www.compassion.com/sponsor_a_child/default.htm [accessed 9 January 2018].

7 Already since the 1970s there had been a rapid expansion of Protestant missions in Haiti as well as in other countries in Latin America. At the same time, this decade was also marked by a mass emigration of Haitians to the United States, fleeing the political persecution and economic hardship of the Duvalier regimes. As evangelism spread throughout the country with the increased presence of American churches and missionary groups, so too did the values of North American capitalist culture and the 'American Dream'. Fred Conway, in his 1978 ethnographic study of rural Haiti, argues that '"missionary Protestantism in Haiti gives rise less to the Protestant ethic of self-help than to the idea that the way to worldly success is identified with direct dependence on the foreign – North American – missionary"'. Signifying modern, capitalist principles, these new congregations have often provided 'career' and self-advancement opportunities in areas where unemployment is high and local prospects are very limited. In the context of Haiti and the seemingly ever-increasing number of NGOs working there (see Part I), it is impossible to dismiss these 'added values' that accompany Gospel preaching and shape religious affiliation among local populations, especially at times of crisis. See Fred Conway, 'Pentecostalism in the context of Haitian religion and health practice' (unpublished doctoral thesis, American University, 1978), p. 193 in Karen Richman, 'Religion at the Epicenter, p. 152.

8 Dan Woolley with Jennifer Schuchmann, *Unshaken: Rising From the Ruins of Hotel Montana* (Grand Rapids, MI: Zondervan, 2011), p. 18. Although the account is Woolley's, Jennifer Schuchmann is listed in the acknowledgements as the co-writer of *Unshaken* and her name appears on the front cover of the book. There is no further information provided anywhere else in the texts on the extent of her involvement in the process of writing the narrative. For this reason, in my analysis I name Woolley as the author and narrator of *Unshaken*.

9 Woolley, p. 17.

10 As Jonathan M. Katz, an American foreign correspondent who was in Haiti during the earthquake observes:

> An enormous effort targeted the collapsed Hotel Montana which had some two hundred people inside – mostly foreigners – when it fell. [...] The places where ordinary Haitians lived and worked – schools, stores, homes, and offices, many with equally ghastly numbers inside – got far less attention. [...] There were many reasons for this disparity. Most foreign rescuers arrived without clear orders where to go. The Haitian government had no reporting mechanism in place for those in need, and there was no formal coordination of rescue efforts [...] Foreign officials knew the UN headquarters, Montana, and Caribbean Supermarket. [...] The coverage of those few featured rescue sites provided a much-needed uplift for viewers abroad. [...] When new rescue teams came in, they knew where to go. They had already seen the priority sites on TV.
>
> Katz, pp. 72–3

11 Woolley, p. 103.

12 Woolley, p. 9.

13 Woolley, p. 9.

14 The title of St Augustine's seminal work means 'both "confessing" in the biblical sense of praising God, and also the avowal of faults'. 'Confessions of St Augustine', in *The Concise Oxford Dictionary of the Christian Church Online*, ed.by Frank Leslie Cross and Elizabeth Anne Livingstone (Oxford University Press, 2005) www.oxfordreference.com/10.1093/oi/authority.20110803095631523 [accessed 3 March 2014].

15 Virginia Lieson Brereton, *From Sin to Salvation: Stories of Women's Conversions, 1800 to the Present* (Bloomington, IN: Indiana University Press, 1991), pp. 3–5 in

Kendrick Oliver, 'How to be (the Author of) Born Again: Charles Colson and the Writing of Conversion in the Age of Evangelicalism', *Religions*, 5 (2014), 886–911 (p. 888).

16 William C. Spengemann and L. R. Lundquist, 'Autobiography and the American Myth', *American Quarterly*, 17 (1965), 501–19 (p. 516), in Charles J. G. Griffin, 'The rhetoric of form in conversion narratives', *Quarterly Journal of Speech*, 76 (1990), 152–63 (p. 152).

17 'confessional', in *Merriam-Webster Dictionary Online* (2015) www.merriam-webster. com/dictionary/confessional [accessed 15 January 2015].

18 'confessional, adj.', in *Oxford English Dictionary Online* (Oxford University Press, 2014) www.oed.com/view/Entry/38782?rskey=mLXXcq&result=2&isAdv anced=false [accessed 13 November 2014].

19 Here, I am drawing on Philippe Lejeune's famous defintion of autobiography as a '[r]écit rétrospectif en prose qu'une personne réelle fait de sa propre existence, lorsqu'elle met l'accent sur sa vie individuelle, en particulier sur l'histoire de sa personnalité'. Philippe Lejeune, *Le Pact autobiographique* (Paris: Éditions du Seuil 1975), p. 14.

20 Stephen A. Shapiro, 'The Dark Continent of Literature: Autobiography', *Comparative Literature Studies*, 5 (1968), 421–54 (p. 435) in Isabel Duran, 'Autobiography', in *The Literary Encyclopedia Online* www.litencyc.com/php/stopics.php?rec=true& UID=1232 [accessed 17 August 2015].

21 Oliver, p. 893.

22 Woolley, p. 81.

23 Woolley, p. 232.

24 Woolley, p. 232.

25 Woolley, p. 232.

26 Woolley, p. 239.

27 Woolley, p. 11.

28 John 11.12. *The Bible: New Revised Standard Version (Anglicised)* (Oxford: Oxford University Press, 1995). [Unless otherwise stated all references come from this edition].

29 Luke 7.11–17.

30 Mark 5.21–43; Matthew 9.18–26; Luke 8.40–56.

31 Woolley, p. 49.

32 'I know your works; you are neither cold nor hot. I wish that you were either cold or hot. So, because you are lukewarm, and neither cold nor hot, I am about to spit you out of my mouth.' Revelation 3.15–17.

33 Woolley, p. 50.

34 Woolley, p. 50.

35 Woolley, p. 49.

36 Kendrick Oliver identifies a number of distinct stages in the formation of the Protestant tradition of conversion narratives in the U.S., from its development in the seventeenth and eighteenth centuries among Reformers in response to the Catholic rite of confession and penance, through an increasing complexity within the genre from the mid-nineteenth century onwards, to a growing realisation of the reconstructive rather than 'authentic' character of conversion narratives and the complex experiences and phenomena they describe. The increasing heterogeneity of the genre also suggests the ambiguous nature and character of experiences of conversion. The tradition of evangelical spiritual autobiography, in Oliver's gloss,

> encouraged the narration of conversion as once-in-a-lifetime event, but it also admitted accounts in which sensations of spiritual resolution were disclosed as premature and superficial and the final attainment of Christian maturity had to await the convert's passage – often interrupted by back-sliding – through a sequence of profound challenges and their claims of new faith.
>
> Kendrick Oliver, 'How to be (the Author of) Born Again', p. 897

37 Peter A. Dorsey, *Sacred Estrangement: The Rhetoric of Conversion in Modern American Autobiography* (University Park, PA: Pennsylvania State University Press, 1993), p. 3.
38 Woolley, p. 51.
39 Woolley, pp. 159–60, 162.

40 'Be quiet! Be quiet!' someone yelled.
    But I wouldn't shut up. I kept yelling. […].
    To *make* them respond to me.
    And then finally, one did.
    'We're not coming for you. We're not going to rescue you! Shut up!'
    [Italics original]. Woolley, p. 163

41 Woolley, p. 123.
42 Mark A. Noll, *American Evangelical Christianity: An Introduction* (Oxford and Malden, MA: Blackwell Publishers, 2001), p. 25.
43 In his analysis of American Evangelical Christianity, Mark A. Noll builds on the earlier work of David Bebbington, who

> has identified the key ingredients of evangelicalism as conversionism (an emphasis on the 'new birth' as a life-changing experience of God), biblicism (a reliance on the Bible as ultimate religious authority), activism (a concern for sharing the faith), and crucicentrism (a focus on Christ's redeeming work on the cross, usually pictured as the only way of salvation.)
>   David Bebbington, *Evangelicalism in Britain: a History from the 1730s to the 1980s* (London: Unwin Hyman 1989), pp. 2–17, in Noll, p. 13

These, according to Noll, are still valid as general characteristics of American evangelicalism, though how this reliance on scripture is expressed or how conversion is understood differs greatly across evangelical churches. Noll, pp. 24–5.
44 Woolley, pp. 224–5.
45 Woolley, p. 231.
46 Woolley, p. 50. Here he is paraphrasing the Parable of the Talents in Matthew 25.14–30.
47 Catherine Wessinger, 'Religious Responses to the Katrina Disaster in New Orleans and the American Gulf Coast', *Journal of Religious Studies* (Japanese Association for Religious Studies), 86 (2012), 53–83 in Elizabeth McAlister, 'Humanitarian Adhocracy, Transnational New Apostolic Missions, and Evangelical Anti-Dependency in a Haitian Refugee Camp', *Nova Religio: The Journal of Alternative and Emergent Religions*, 16 (2013), 11–34 (p. 24).
48 One source for this vision of history is a literal interpretation of Deuteronomy 30.15–20:

> See, I have set before you today life and prosperity, death and adversity. If you obey the commandments of the Lord your God[b]. that I am commanding you today, by loving the Lord your God, walking in his ways, and observing his commandments, decrees, and ordinances, then you shall live and become numerous, and the Lord your God will bless you in the land that you are entering to possess. But if your heart turns away and you do not hear, but are led astray to bow down to other gods and serve them, I declare to you today that you shall perish; you shall not live long in the land that you are crossing the Jordan to enter and possess. I call heaven and earth to witness against you today that I have set before you life and death, blessings and curses. Choose life so that you and your descendants may live, loving the Lord your God, obeying him, and holding fast to him; for that means life to you and length of days, so that you may live in the land that the Lord swore to give to your ancestors, to Abraham, to Isaac, and to Jacob.

49 McAlister, p. 14.

50 The most recent General Census of Population and Living Conditions (2003) conducted by L'Institut Haotien de Statistique et d'Informatique (The Haitian Institute of
Statistics and Data) indicates that 54.7 per cent of population declares itself Catholic,
15.4 per cent Baptist, 7.9 per cent Pentecostal and approximately 10.2 per cent
declares no religious affiliation. L'Institut Haïtien de Statistique et d'Informatique,
*Recensement Général de la population et de l'Habitat: Enquête nationale sur la
population sa structure et ses caractıristiques dımographiques et socio-ıconomiques*
(L'Institut Haïtien de Statistique et d'Informatique 2007) www.ihsi.ht/rgph_resultat_
ensemble_population.htm [accessed 17 February 2015].

51 Lukeson is a Haitian man who is trapped with Woolley under the rubble of Hotel
Montana, where he worked prior to the earthquake. He is then rescued by the same
emergency teams. Throughout the text, the narrator uses stylised speech (e.g. grammar
mistakes, 'dat' for 'that') for Lukeson, which is meant to mirror his relatively poor
command of English. Woolley, p. 110.

52 'The underworld; the abode of the dead or departed spirits, conceived by the Hebrews
as a subterranean region clothed in thick darkness, return from which is impossible.'
'sheol, n.', in *Oxford English Dictionary Online* (Oxford University Press, 2014)
www.oed.com/view/Entry/177962?redirectedFrom=sheol [accessed 3 March 2014].

53 Richman, pp. 152–3.

54 Richman, p. 154.

55 In psychological terms, the idea of disaster as divine punishment and victims'
responsibility for their own suffering is termed ' "negative religious coping" '. Harold
G. Koening, *In the Wake of Disaster: Religious Responses to Terrorism and
Catastrophe* (Philadelphia, PA: Templeton Foundation Press, 2006), pp. 31–3, in
McAlister, p. 24.

56 For example, Kanès, one of the members of the Assembly of God congregation inter-
viewed by Richman, provides the following commentary on the post-¬earthquake
religious landscape in Ti Rivyè, Léogâne:

> A lot of people converted after the goudou-goudou. They were afraid. They had
> never experienced anything like it. The earth opened up and then it rose and fell.
> People thought it was Bondye (not lwa). It was a natural occurrence like the hurri-
> canes that come every year. They believed Bondye wanted them to convert; they
> thought if they did, it would protect them the next time. But many have already
> left it. They weren't good conversions.
>
>   Richman, p. 158

9   Imagining novel lives

In contradiction to Woolley's autobiographical account of his double survival and the journey of spiritual renewal it inspired, Nick Lake's *In Darkness* and Laura Rose Wagner's *Hold Tight, Don't Let Go* are young adult novels which imagine, with varying levels of realism, the experience of surviving the disaster and remaking one's life in its immediate and long-term aftermath. Lake's award-winning book, the first novel of the British author who works in publishing, lives in Oxfordshire (UK) and watched the 2010 disaster from afar,[1] presents the reader with a story of a young boy, Shorty, who is trapped under the rubble of one of Port-au-Prince's hospitals.[2] He was treated there a few days prior to the disaster following a shooting between rival gangs in Cité Soleil, one of the Haitian capital's most deprived neighbourhoods. *In Darkness* is dedicated to 'the people of Site Solèy' (Kreyòl for Cité Soleil) and opens with two epigraphs. The first is an extract from Toussaint L'Ouverture's[3] letter to Napoléon Bonaparte, and the second consists of the last six lines from William Wordsworth's 'To Toussaint L'Ouverture' (1802).[4] Although initially the link between the dedication, the epigraphs relating to the Haitian Revolution, and the subject of the novel (the 2010 earthquake) seems rather unclear, the rationale behind this pairing soon becomes apparent. The novel is divided into two sections, 'Now' and 'Then', and intertwines the story of Shorty with that of Toussaint L'Ouverture, the famous leader of Haiti's struggle for independence. The epigraphs act as framing devices, foreshadowing the text's narrative structure, which, at once, joins and switches between the fictional story of Shorty and the fictionalised account of Toussaint L'Ouverture's life and the beginnings of the Haitian Revolution. The alternating design of the text allows Lake to create points of comparison between Shorty and Toussaint, to explore the complex significance of these two events – each being a historical marker in its own right – within the conventions of the genre and, in turn, to suggest ways in which individual/collective subjectivity, rescue and recovery can be envisaged in the wake of such testing moments.

Laura Rose Wagner's *Hold Tight, Don't Let Go*, for its part, takes the form of a fictional diary of a young Haitian girl, Magdalie, who lives with her aunt and cousin, Nadine, in Port-au-Prince. The two inseparable cousins, who consider themselves as twins, 'better and more powerful for being together',[5] both survive the January earthquake and try to find themselves within the new reality of the devastated capital and the forced displacement. However, Nadine's family in the United States wants her to leave the ruined city and join them abroad. The beloved cousin's departure from Haiti is a second blow for Magdalie: healing and recovery seem all the more unreachable without her 'other-half'. The novel, partly based on the author's own experience of surviving the tremors and published a few years later than Lake's, starts with the day of the earthquake and follows Magdalie's struggles and her attempts to build a new life for herself in Haiti, as recorded in her diary between 12 January 2010 and October 2011. The form privileges an intimate insight into the daily life and Magdalie's determined quotidian efforts to find her way among the visible destruction and the emptiness she experiences after her cousin's departure. This sense of intimacy between the reader and the text are further reaffirmed in the novel's dedication, to Melise Rivien who died on 12 January, as well as Wagner's biographical information provided in the book's jacket: the author, 'like countless others, survived the January 2010 earthquake through the grace of ordinary people in Port-au-Prince'.[6] A narrative of the protagonist's, and possibly the author's, journey towards healing, the novel captures the most difficult challenge of post-earthquake recovery and remaking: holding tight and not letting go, defiantly, day in, day out.

The genre of the young adult novel distinguishes *In Darkness* and *Hold Tight, Don't Let Go* from other accounts analysed earlier in this book and allows for a productive consideration of insights offered by these creative responses into the experience of living through the disaster and its immediate and long-term aftermath. The category of young adult literature (YAL) is a complex one and encompasses a wide selection of books targeted primarily at the 13–18 age range. As a distinct and targeted designation, YAL originated in the United States as an area of 'children's literature addressed to the adolescent/teenage market', and has since been adopted in other countries in response to the growth of the juvenile reading audience.[7] At the most basic level, YAL could be defined as '(1) that written especially for them [young adults], and (2) that which, while not written especially for them, is available for their use'.[8] Yet even this broad explanation, used in an early textbook and criticised by Jannice M. Alberghene in 1985, does not provide firm or clear-cut criteria by which to classify and evaluate YAL – an increasingly elusive and complex category. A decade after Alberghene's attempts to characterise the term, Michel Cart remarked, with some discernible irritation, that 'even to try to *define* the phrase "young adult (or adolescent) literature" can be migraine inducing' [italics original].[9] At the same time, this conceptual challenge points to the unavoidable blurring of lines between literature written for children and that aimed at young adults, an obscuring that 'reflects the lack of absolute boundaries between childhood, adolescence, and adulthood'.[10] Rather than trying to simplify this, at times, mind-boggling heterogeneity, recent discussions of YAL seek 'new paths of analysis',[11] hoping to move beyond questions of thematic relevance, pedagogical dimension and accessibility that were previously hailed as the distinct and, at times, the only qualities of the genre and the main reasons for the inclusion of YAL in school curricula.

Whereas these characteristics can still be found in Lake's and Wagner's novels, they are only a part, not the defining features, of the texts' complex fabric. Rather, both *In Darkness* and *Hold Tight!* demonstrate that 'literature for adolescents might be stylistically complex, that it might withstand rigorous critical scrutiny, and that it might set forth thoughtful social and political commentaries',[12] directly challenging earlier approaches that saw YAL as an aesthetically less developed form. Formally and conceptually nuanced, YAL, like other narrative genres, is shaped and transformed 'by topics and themes that years ago would have never ever been conceived'[13] such as biotechnology, post-humanism and, not least, environmental hazards. Building on this recognition of the genre's complexity and sophistication, 30 years after Alberghene's discussion of YAL, Stephen Roxburgh echoes her point and accentuates the artistic qualities YAL and other literary forms share: *'There is no difference* between the young adult novel and the adult novel. There are distinctions to be made between them, but they are not different art forms' [italics original].[14] Rather, in accounting for the specificity of the genre, he argues that narrative point of view and, even more so, voice, are 'the very essence of the young adult novel'.[15] The character 'is made manifest in and by the protagonist's voice'[16] rather than through action, appearance or description. In short, '[v]oice *is* character *is* plot' [italics original].[17] Precisely, in their use of and search for a distinct voice for the experience of the 2010 disaster, Lake and Wagner respectively demonstrate the novels' capacity to engage with complex questions of identity formation and socio-political issues: terrorism, immigration or disasters such as the 2010 earthquake.[18] At the same time, as they craft an imaginative response, these texts attempt to give a voice and a privileged position to children's experience of the disaster – a perspective largely omitted elsewhere.

In considering these literary modes of representation and their use of realism, or the interweaving of magical realist elements, and the significance these aesthetic components have for an understanding of the 2010 earthquake, it is important to recognise that factual accuracy is not the main goal for the two texts, nor is it a fixed characteristic of their genre. Throughout *In Darkness*, for example, the historical narrative of Toussaint L'Ouverture functions as a point of departure for an unorthodox creative engagement with the country's recent and more remote past:

> I occasionally simplified and adjusted the facts to fit into the shape of the story [...] The simple answer is that I believe that the book is true in essence [...] I did not invent the character of Toussaint l'Ouverture [*sic*], and I have been faithful to his story, at least in spirit and in essentials. It was necessary to smooth out the history to some extent.[19]

In this extended passage from the author's note, the motivations guiding the structure of the text are spelled out clearly: the novel wants to alert the reader to key contemporary and historical issues without claiming to provide an authoritative explanation of either defining events or the figures that symbolise them. Instead, the account, 'true in essence', weaves between fictional and factual material, and both of its main narratives seem equally important. The author's self-reflexive comments, as well as the book's target audience, seem to pre-empt the twin charge of uncritical appropriation of the Haitian Revolution and sensationalisation of the traumatic experience of the 2010 earthquake. At the same time, both factors point to the ways in which each narrative account, regardless of the extent of its realism and factual accuracy, represents a 'smoothing out' of historical complexities.

Rather than attempting to convey the intricacies of the late eighteenth-century Revolution or the early twenty-first-century earthquake, *In Darkness* prioritises an immersive, affective experience for the reader that might then inspire him/her to reach out to additional resources and to learn about the historical background to these two epochal events. Such complementary material is hinted at in the publisher's closing note: 'For more information about Nick Lake and his astonishing novel, including an author interview and a reading guide, visit www.in-darkness.org.'[20] The dedicated website (recently discontinued as a stand-alone page), gave, among other things, a brief overview of the key figures and events from the book. The page also had a list of recommended reading on Haiti and the slave trade, a downloadable reading guide and even an indirect fundraising appeal.[21] In contrast to other discontinued supporting material, the reading guide is still available in an updated and abridged form from the publisher's (Bloomsbury) book page and consists of a brief summary of the text. It also provides a number of extracts from the novel, followed by a list of suggested questions to related topics such as those of power, healing, religion and spirituality, which can be easily discussed in a variety of classroom contexts and which vary significantly in terms of the complexity and level of knowledge they assume.

This sense of analogy and necessary adaptation is central to the book's design and the generic conventions it negotiates. At the most basic level, the bipartite structure of the text already hints at the comparability of the Haitian Revolution and the 2010 earthquake in terms of their symbolic impact. The formal correspondence, later complemented by the thematic interplay between the 'Now' and 'Then' sections of the text, also suggests the ways in which the adventures and lessons learned by the two protagonists are applicable to the reader's everyday experience. This last aspect of *In Darkness* is particularly important in the context of the book's designation as a young adult novel: through a creative rendering of key events in Haiti's history it popularises the country's past and offers a compelling story that frames the book's more pedagogic message. The acknowledged simplification of the historical account does not limit the 'success' of the novel. Rather, such abridged presentation and the double parallel enable the narrator to create a formally complete and generically consistent narrative that speaks to its target audience in the ways that are deemed most relevant to it.

Similarly, *Hold Tight!* includes additional material – a glossary of Haitian Kreyòl terms, a brief history of Haiti and a list of suggested reading – which widens the appeal of the book, aims to help the readers appreciate and contextualise the events of the plot, the earthquake but also the cholera epidemic. In so doing, the novel seeks to challenge common portrayals of Haiti as 'a poor country, populated mostly by dark-skinned people[,] a place of suffering, violence, and disaster [...] that needs help'.[22] There is a clear critical dimension to these supplementary resources: they hope to inspire in readers a sense of awareness and inquisitive curiosity about Haiti that would go beyond voyeuristic pity, compassion fatigue or disinterested ignorance. Available as an appendix to the novel rather than as a separate online resource, the supplementary material has a much wider reach and potential impact and facilitates a more complex engagement with the text, adding to the sense of accuracy of its portrayal of everyday life in Haiti. At the same time, similarly to *In Darkness*, these extratextual features in *Hold Tight!* raise the question of the novel's realism and its function. Here, too, the aim is not to present a historical reconstruction of events but rather to capture the 'emotional truth' of the experience of the earthquake and its aftermath, 'a "true" story' of these days, 'meaning a story that expressed some kind of emotional truth and that was true to experience'.[23] Together, the accompanying material and the authorial comment echo the narrative's aim to complicate the reader's perception of the 2010 disaster, complement as well as destabilise the 'truthfulness' and comprehensiveness of seemingly exhaustive accuracy of non-fiction and empirical depictions and analyses of the January tremors.

In addition to its role as a potential educational support tool, the extratextual resources add an unexpected dimension to both books, communicating their ambition to contribute to post-earthquake reconstruction in direct and tangible ways – an impulse also clearly expressed in Woolley's text. For example, the earlier, dedicated web page for Lake's novel had a separate section, entitled 'Haiti Aid', which emphasised the importance of sustained relief effort:

> Charities are still continuing their aid in Haiti by helping people recover from the 2010 earthquake. Ten months after the earthquake, an outbreak of cholera added to Haiti's woes. To find out more about aid in Haiti, and how you can help, visit: Disaster Emergency Committee, Save the Children, Red Cross.[24]

Here, this paratextual comment modifies somewhat the book's more critical narrative representation of foreign aid and its impact on Haiti. In particular, the novel's portrayal of the romance between a white female relief worker and Dread Wilmè, the leader of one of the gangs, might indeed be treated as a pointed critique of naïve humanitarian efforts to 'help' Haiti without addressing the structural issues that are at the source of the country's vulnerability. In direct tension to this thematic critique, the direct recommendation, most likely targeted at the teachers who might be consulting the supplementary material, clearly points to the book's ambition to go beyond its generic category of YAL and reach a wider readership as well as help those directly affected by the disaster. Yet, in a paradoxical reversal, the author's initiative is halted half-way. Discontinued from the publisher's page, the absent aid section – now deemed obsolete

– tellingly reveals the limitations of timed, fixed-term relief efforts. *Hold Tight!*, for its part, takes this commitment to contribute to recovery efforts in tangible ways one step further with the book itself, rather than the reader's later donation inspired by the narrative, being a form of donation. As the brief information on the book jacket informs us, '[a] portion of the author's proceeds from the sale of this book goes to the Centre d'Education Spéciale [...] and Ti Kay Haiti [...] in support of their work providing education and healthcare in Haiti'.[25] The authorial decision to support Ti Kay, a special needs education centre, not only adds a sense of urgency to the account but also is a clear demonstration of the ways in which YAL can be a site of a committed ethical-political intervention.

At the same time, while reinforcing the respective aims of the three texts, their form and extratextual features also expose the limitations of their narrativisation of disaster. The need to provide extensive additional material, such as author's notes and online guides, so as to frame the autobiographical conversion account as well as contextualise the fictional narratives, seems to imply that narrative on its own struggles to engage sufficiently with the lasting impact of the disaster. Self-reflexive authorial commentary, as well as the more or less explicit fundraising appeal of each text, complement the work of the narrative and respond to the need for a tangible translation of the affective experience of reading. Finally, despite key generic differences, these paratextual appeals create surprising parallels between Woolley's autobiographical account and Lake's and Wagner's fictional works: they all express, albeit in different ways, a sense of responsibility for and towards the wider community of earthquake survivors beyond the realm of narrative representation.

Twinned histories, twinned lives

The formal conventions and aspirations of Nick Lake's *In Darkness*, discussed above, raise a key question on the use of narrative voice in the text and the relationship it sets up between narrator, narratee and audience. The two narrative strands, one narrated by Shorty and the other by Toussaint L'Ouverture, are distinct. The latter is a third-person omniscient narration which recounts the life of Toussaint L'Ouverture from his ascent to power to his imprisonment by the French colonial authorities. The former is a first-person narrative addressed to an imagined 'you', the recipient of Shorty's account, and has a strong dialogic quality: the teenage boy is looking for the narratee's understanding and a certain degree of sympathy. Shorty is at once an autodiegetic and intradiegetic narrator: he tells his own story and takes part in it.[26] His account is narrated consistently in the present tense; the narrative unfolds in time with the reader and the narrator being equally uncertain of its outcome. The novel's opening, *in media res*, draws the reader immediately into the story and the unpredictable world of darkness and dreams:[27]

> I am the voice in the dark, calling out for your help.
>   I am the quiet voice that you hope will not turn to silence, the voice you want to keep hearing cos it means someone is still alive. I am the voice

calling for you to come and dig me out. I am the voice in the dark, asking you to unbury me, to bring me from the grave out into the light, like a zombie.[28]

This captivating opening address necessitates a response: someone is calling out for help and, as the likely recipients of this cry, we, the readers, cannot turn away. The dramatic framing also establishes an intimate relationship between the narratee, who can be defined as 'a figure imagined within the text as listening to – or receiving a written narration from – the narrator',[29] and the unnamed voice in the dark. Shorty calls out to this unspecified narratee repeatedly throughout the account with the narratee's presumed yet unvoiced questions shaping the narrator's recounting of events. For example, as he is telling the story of his birth and early years, one of many occasions in the text when Shorty draws on different sources such as Aristide's book or his mother's memories, the narrator suddenly interrupts his account to ask: 'You're thinking, how can I know this? How can my manman have remembered Aristide's words? And I answer you – she didn't. But Aristide wrote them down in a book, and I have that book still.'[30] This and other recollections and self-reflexive interjections are incorporated into the narrative as seemingly unmodified fragments of recalled conversations. As such, they add to the complex fabric of the text, revealing the narrator's emotional maturity beyond his years that increases the credibility of his narration and his later moral evaluation of his own actions.[31]

In contradistinction, the second thread of the novel, which focuses on Toussaint L'Ouverture, is a more conventional third-person narration, with the omniscient narrator recalling events directly preceding the outbreak of the Haitian Revolution. The future leader is also presented as an outsider: his rational attitude and philosophical sophistication distinguish him from his comrades. For example, during the Bois Caïman Vodou ceremony – the symbolic initiation of the Revolution under L'Ouverture's leadership – L'Ouverture's sceptical and analytical reserve is in stark contrast to the animated cries of his fellow countrymen who, as he sees it, are merely participating in some form of folkloric superstition.[32] Although it is never revealed who the narrator of the historical narrative is, as the novel develops it seems increasingly possible that Toussaint's story unfolds in Shorty's dreams. Indeed, the transitions between the two sections, 'Now' and 'Then', often occur when Shorty seems to be falling asleep or when Toussaint is dreaming.

This sense of uncertainty of the boundary between the real, the imagined and the magical elements, which is a characteristic of magical realist aesthetic, introduces a further level of complexity in the novel's exploration of ideas of subjectivity, agency and recovery. The young protagonist, no longer a child but not yet an adult and unsure of what is real and what is merely a dream, occupies a liminal space in which magic acts as a catalyst for Shorty's subjective transformation as well as the young adult reader's socialisation.[33] On other occasions, however, an image, theme or prolepsis bridges the two strands; for example, both Shorty and Toussaint devise a cooling system in order to keep food fresh for longer.[34] Together, the repeated parallels and the increasing blurring of boundaries between the two story threads introduce an elaborate interplay between the two narrative strands, suggesting that notions of post-disaster sub-jectivity, agency and recovery are similarly envisaged as necessarily grounded in the country's revolutionary past, which, once again, becomes a foundational point of reference and a new beginning.

The formal parallels the novel establishes between pre-revolutionary and post-earthquake Haiti emphasise similarities between the two protagonists, Shorty and Toussaint L'Ouverture, with events from the imaginatively invoked Revolution foreshadowing specific moments in Shorty's life in contemporary Haiti. *In Dark-ness*, like *Unshaken*, traces the movement from fragmentation to wholeness and the accompanying notions of subjectivity formation and personal recovery. However, Lake's novel considers them through the lens of Vodou thought and symbolism by making extensive references to *marasa* (divine twins) and via the metaphorical use of the figure of *zonbi* and zombification.[35] As such, *In Darkness* distinguishes itself from the other texts analysed here by its explicit engagement with Vodou symbolism and Haiti's national history, even if at times, as I go on to demonstrate, the more culturally-attentive engagement risks being overshadowed by the rehearsed, widely circulated pop-culture misunderstandings of Vodou. Also, somewhat akin to *Haïti, kenbe la!*, *In Darkness* seeks to locate the significance of the 2010 earthquake, the violent break it inaugurates and the ongoing violences, through which its aftermath is made visible, by turning to the country's past and interweaving the story of the earthquake with that of the Haitian Revolution.

By invoking the story of the late eighteenth-century Revolution, which has been the subject of numerous creative works[36] and 'has always been an ideo-logical battleground',[37] the novel suggests the need for a reconsideration of found-ing national narratives and their relevance to the shattered post-earthquake context. In effect, *In Darkness* brings together two familiar discourses on Haiti's predicament, the country's unique historical beginnings, and its equally excep-tional contemporary impoverishment, and ponders the possibility of contributing to new narratives in the aftermath of the disaster. Yet rather than suggesting an easy comparability and complementarity of these two historical processes, one that would suggest a cyclical view of Haiti's history, the novel seeks to envisage a post-earthquake transformation that would take up, restore and make audible again the revolutionary call for an equal tomorrow.

The figures of *zonbi* and *marasa*, which are used to complement the formal and thematic twinning of these two narrative strands, have a significantly different spiritual and historical-cultural resonance, however – a discrepancy that carries far-reaching ethical implications for Lake's narrative. Whereas the divine twins and the cult of twins are central to Vodou, the figure of *zonbi*, an important trope in Haitian folklore, has long been appropriated and misused in Western popular culture to present a highly racialised and denigrating view of Haiti, its history and culture. From the early days of colonisation, Vodou held the atten-tion of colonial travellers, whose tales offered an image of the religion as a form of satanic ritual combining cannibalism, superstition, the cult of the dead and human sacrifice.[38] Racist ideology, on the rise in the nineteenth century, further sustained this view of the country as trapped in a satanic-like darkness that could only be countered by emulating the enlightened practices of modern European states. A century later, during the American Occupation (1915–34) in Haiti, similar discourses of the country's cultural 'backwardness', which linked race, religion and governance, were employed to frame the invasion and to justify, during that time, such military actions 'as pillaging the voodoo temples and destroying the "idols" of the African ancestors'.[39] Claiming to provide readers with an unprecedented insight into this dark, magical and unchartered '"land of the ghosts'",[40] many of the texts published in that period, William Seabrook's infamous *Magic Island* (1929) among others, effectively only solidified earlier racialised misperceptions of Vodou, which framed Haitians as the Other and Haiti as an impenetrable, exotic land. These radically different histories of misuse of the two cultural-spiritual tropes resound throughout Lake's narrative.

After these early publications, subsequent literary and cinematic representations quickly appropriated the trope of the *zonbi*, which came to signify and at once confirm all that is irrational and sensational in Vodou, permanently altering cultural understandings of what a *zonbi* is. Film productions such as Victor Halperin's *White Zombie* (1932) and Jacques Tourneur's *I Walked with a Zombie* (1943) are early examples of what is now classified as a distinct subgenre of horror movies, 'zombie films'.[41] In contemporary culture – films,[42] literature, TV series, 'zombie-walks' and flash-mobs, video games,[43] even financial analysis[44] – this image of the *zonbi* as a flesh-eating, contagious, soulless corpse, a walking-dead, is predominant. These cultural representations, along with pseudo-scientific studies of 'zombie death', the most famous being Wade Davis's *The Serpent and the Rainbow* (1985),[45] not only validated popular cultural misconceptions, but, years later, were also a direct source of those anti-Haitian attitudes that, to cite a particularly scandalous example, erroneously deemed Haitians responsible for the spread of AIDS in the United States.[46] In both popular media and scientific circles, Haitians were presented as disease-ridden, backward and at risk of contracting or spreading AIDS.[47] Yet, as research has shown, AIDS did not come to America with Haitian immigrants at all; rather 'it [went] south with North American tourists'.[48] In a racist extension of the liminal metaphor, new Haitian immigrants disembarking at America's shores, always seen as potential carriers and thus as a threat, were *zonbi*-like figures 'leaving a trail of unwinding gauze bandages and rotting flesh'.[49] These appropriations and misrepresentations of Vodou and 'zombie mythology', which contrary to its spiritual significance in Vodou was presented only in terms of a reawakened soulless dead body, a zombie, increasingly featured in popular and literary discourses, becoming a synonym for Haiti and its culture that has endured to the present day.

Many of the post-earthquake media responses to the 2010 disaster have been quick to evoke these facile images of Haiti's barbarity, backwardness and underdevelopment as a direct consequence of its dark 'voodoo' past (as epitomised in the Bois Caïman ceremony). Pat Robertson's 2010 reading of the earthquake is one such extreme example of the strategic abuse of Vodou and its cultural importance for the Haitian Revolution. Robertson's simple logic of equating all Vodou with evil, which imposes on Vodou Protestantism's dualistic theological system' (Satan as the enemy of God),[50] leads him to a damning verdict: that the start of the Revolution and the founding of Haiti are rooted in this diabolical practice, and that Haitians now need to repent for this zombie-like past. Yet even seemingly more nuanced readings, like those offered by the *New York Times* (January 2010) under the guise of objective market analysis, claim that Vodou is responsible for Haiti's 'progress-resistant' culture.[51] For example, David Brooks's commentary in the *New York Times*, which was quickly followed by Lawrence Harrison's equally disparaging claims in the *Wall Street Journal* (February 2010), rehearse and perpetuate denigrating portrayals of Vodou spirituality, seeing it, alongside with Haiti's culture, as the only reason for 'Haiti's unending tragedy'.[52] Such pronouncements are ultimately not too dissimilar from Robertson's claims. Both present the country's history, heritage and traditions as inherently inferior, implying that these practices hinder Haiti's cultural, economic and political 'progress', which is defined in terms of the abandonment of non-Western customs, the expansion of the neoliberal market economy, and Haiti's increased participation in global markets.[53]

In effect, these inaccurate accusations, criticised by Benjamin Hebblethwaite among others, fail to provide any meaningful insight into the real cases of Haiti's underdevelopment and reveal 'stereotypical and xenophobic rhetorical tropes from a received tradition, one connected to impulses stemming from colonial period in which a culture of racism, exclusion, and slavery inflicted much harm on African religions'.[54] Reductive and politically motivated readings of Vodou avoid engaging with the philosophical complexity of this very different cosmology, presenting it instead as the sole reason for Haiti's socio-economic difficulties. Finally, within such rehearsed and repeated interpretations the 2010 earthquake and its aftermath is just another fatal manifestation and confirmation of Haiti's inability to govern itself.

There is little connection between this imagined 'voodoo', epitomised in 'flesh-eating zombies', and the real Vodou tradition; rather what is 'religious about the spirit *zonbi* – the fact that the human spirit lives on beyond death in an invisible part of the cosmos and has dealings with the living – has been turned inside out'.[55] The *zonbi* as 'a former human with a body but no soul, spirit, consciousness, interiority, or identity' has 'become the common understanding of zombies in contemporary culture'.[56] These simplistic appropriations of the *zonbi* and zombification as a trope ignore their complex historical roots, rich folkloric genealogy and spiritual significance as well as their use in Haitian literary tradition. First, the historical genesis of belief in zombification cannot be separated from Haiti's African origins and the slave system of St Domingue in the eighteenth century, since 'the zombie is the living-dead, subjected completely to the wills and caprices of a master (his proprietor) for whom he must work'.[57] The figure of the *zonbi* is 'the ultimate sign of loss and dispossession'[58] as well as 'the perfect realisation of the slave condition, the very ideal sought by the master in his slave'.[59] With the 'memories of servitude [being] transposed into a new idiom that both reproduces and dismantles a twentieth-century history of forced labor and denigration that became particularly acute during the American occupation of Haiti'.[60] From the 'body-and-soul-fracturing dictatorships'[61] of François 'Papa Doc' and Jean-Claude 'Baby Doc' Duvalier (1956–71 and 1971–86), onwards, zombification as a literary and visual trope has represented the broader experience of exploitation, alienation, suffering and victimhood which, albeit in different forms, is shared by equally disenfranchised social groups, whether members of the political opposition or the urban poor.[62] *Zonbis*, seen as the antithesis of the romanticised hero in the Indigenist novels[63] 'and of the noble peasant extolled in Indigenist theoretical writings',[64] stand for all that is 'uncontainable within an order of things';[65] as creatures trapped in the present and subject to the whims of black magic, they provide a 'negative mirror of what is or should be the human self'.[66] These are not the bloodthirsty monsters of the movie screens, but rather the 'lowest being[s] on the social scale [...] reduced to [their] productive capacity'.[67] In short, *zonbis* have no agency and zombified individuals lose their willpower, becoming fully obedient to their master.

In Haitian folklore, contrary to the image of the zombie in Western films, the *zonbi*[68] can only become violent if ordered by his master, does not eat human flesh and is not contagious.[69] Second, zombification in the Vodou tradition refers to the act of capturing one of the two souls that each human being possesses. The first one is 'the little good angel, *ti bon anj* (responsible for the will) and the [second is] big good angel *gwo bon anj* (responsible for consciousness)'.[70] For Vodouists then, 'the act of zombification is a matter of capturing the *ti bon anj* and thus exerting an absolute power over the individual'.[71] In Afro-Haitian religious thought, Cartesian mind-body dualism is replaced by a belief that 'part of [an individual's] spirit goes immediately to God after death, while another part lingers near the grave for a time [and it] is this portion of the spirit that can be captured and made to work'.[72] However, this immobilised subjugated state lasts only as long as the creature is denied salted food, and as soon as it tastes any salt it awakens and becomes aware of its predicament. Dezombification therefore denotes individual/collective awakening and reassertion of one's agency often leading to action or, in political terms, some sort of affirmative positioning akin to historical practices of resistance such as *marronage*.[73] As such, contrary to popular misrepresentation, zombie's subjugation, and by extension the social exploitation it metaphorically represents, is not necessarily a definitive one, and there is always a possibility of awakening; in Kaiama L. Glover's words: 'the hero is always dormant in the zombie'.[74] Also employed as a 'metaphor for the enslaved, for the socially alienated, for the politically disenfranchised',[75] *zonbi* 'survives as the remnant of loss and dispossession'.[76] The complexity of the *zonbi* figure and its historical significance contributed to its importance for the collective imaginary as well as its continued relevance.

Alongside the *zonbi*, *marasa*, the divine twins, are an equally complex symbol, firmly rooted in the Vodou religion that both Lake and Wagner employ in their respective novels to consider transformations and modifications of self and community after the earthquake. As distinct from zombies, the twins have received a much smaller degree of popular Western attention and have not been appropriated to the same extent by contemporary cultural producers. The twins continue to be of great importance to Vodou practitioners and occupy an unchallenged position within Vodou's divine pantheon. The spiritual and cultural significance of *marasa* can be traced back to Africa, more specifically Dahomey (present-day Republic of Benin),[77] where 'twins were considered sacred'.[78] A simple set of twins, which can be of different genders, is referred to as *marasa de* whereas a set of three is called *marasa twa*. *Dosou* denotes a male child born after a set of twins, whereas *dosa* signifies a female. The child complements the twins and is seen as being even more powerful than the twins themselves.[79] In the Vodou-inspired worldview, the Divine Twins are a whole and a three.[80] They contain, in Maya Deren's gloss, 'the notion of the segmentation of some original cosmic totality' and are 'a celebration of man's twinned nature: half matter, half metaphysical; half mortal, half immortal; half human, half divine'.[81] They are at once first humans and first ancestors: they are the origin of *loa* (also spelled as *lwa*)[82] (Vodou spirits) and, as such, are the first to be saluted at the start of all Vodou ceremonies. *Marasa* symbolise 'abundance, plurality, wholeness, innocence, and newness, thus their representation as children'.[83] Sometimes referred to as *kalfou marasa* (Haitian Kreyòl for intersection or junction), they are also considered to be 'guardians in control of the crossroads and [...] are linked to the Lwa-spirit Legba, who opens the gate to the crossroads'.[84] This continued metaphysical importance of *marasa* results in a special regard for twins, who are also seen as two parts of a whole: whatever affects the one (e.g. violence or disease) threatens to affect the other and if the two *marasas* are not treated well, then the third twin restores justice.[85]

It is in this double context of the historical-spiritual origins of the *zonbi*, its appropriation and misuse as zombie, and the *marasa* and its contemporary translations that Lake's *In Darkness* use of this trope as well as the figure of the divine twins has to be considered. Of course, the target audience of the text, young adult readers, and the aims of the novel greatly influence the narrative's ability to engage fully with these complexities. However, it is precisely cinematic and literary representations – often relying on repetition, sensationalisation and high drama that borders on the pornographic gaze – that largely gave rise to and continue to sustain these stereotypes of Vodou and Haiti. As such, the importance of imagery and metaphor, and their historical-political resonance in works like Lake's, cannot be overlooked – an issue central to my subsequent analysis of the text's visions of collective recovery.

In Darkness interweaves and connects these two figures of zombie and *marasa* to trace Shorty's and Toussaint's respective transformations, which are triggered by very different yet equally powerful experiences of exclusion and violence. Whereas *marasa* is a key metaphor for Shorty's changing sense of subjectivity, the figure of the zombie, as used in the text, suggests a possible awakening of a new revived self and collectivity. After losing his twin sister, his 'other half', Shorty considers himself a 'half-person'[86] and goes on to become a full member, a 'vre chimère' of the local gang, Route 9.[87] As an inhabitant of a 'no-man's land'[88] in Cité Soleil, a cordoned-off shanty town in Port-au-Prince's metropolitan area, and as a gangster, he is a ghostly figure trying to formulate, through violence, a new 'unbroken' identity. Shorty hopes that by fighting the rival gang, which he thinks is responsible for his sister's disappearance, he can avenge his loss, give a meaning to the void he is experiencing, and find a new, 'whole' self. The acts of violence he needs to commit in order to be accepted into the gang and rise within its ranks are meant to confirm his newly earned 'full' gangster status and counter the fear of a permanent subjective emptiness:

> When you keep hurting someone, you do one of three things. Either you fill them up with hate, and they destroy everything around them. Or you fill them up with sadness, and they destroy themselves. Or you fill them up with justice, and they try to destroy everything that's bad and cruel in this world.
> Me, I was the first kind of person.[89]

As he is lying underneath the ruins of the hospital, in a moment of retrospective moral self-evaluation, Shorty comes to acknowledge the reasons for his previous actions and his eventual transformation: they were responses to the violence inflicted upon him as well as his negative experience of double marginalisation. Having first tried to combat this dual exclusion before the earthquake struck, now battling to stay alive, Shorty is fighting against a very different 'double death': that of not being found in time and that of being forgotten without being forgiven.[90] Desperate, he calls out to anyone and everyone who might hear him, hoping, against all odds, for an opportunity to explain and justify his life of crime, even if only to an unnamed addressee: 'Maybe, maybe, if I tell you my story, then you'll understand me better and the things I've done. Maybe you'll, I don't know, maybe you'll … forgive me.'[91] This hopeful and defiant act of telling a story – a full life-story recounted, for once, from Shorty's perspective – becomes then an attempt for the protagonist to reclaim ownership and write a different ending to his life, one of a 'full person', not a *chimère*, a violent ghost.

Violence, the main driving force and defining component of Shorty's fractioned self also shapes – albeit in very different ways – Toussaint's transformation from a coachman to a revolutionary leader:

> [...] there were three kinds of slaves, three kinds of people. There were those who were so filled with hate by their experience, by their oppression, that they snapped and destroyed property or people. There were those who were so filled with sadness by their experience that they snapped and destroyed themselves [...]. The third kind of person, though, was filled by their experience with a fierce desire to make things right in the world, to redress the balance.

In the darkness, Toussaint fancied that he was the third kind of person, and to fire his soul, to fill himself with a sense of the need for justice, he called up the faces that embodied for him slavery's evil.[92]

The extended quotation establishes a clear correspondence, by citation and repetition, between Shorty and the fictionalised Toussaint who, within the young narrator's characterisation, is the third type of person: in response to all-surrounding violence and hurt, he tries 'to destroy everything that's bad and cruel in this world'. Thanks to the focalisation and high level of self-reflexivity of the passage, the historical figure is brought alive and rendered more relevant to the contemporary reader. L'Ouverture is not some distant revolutionary general but rather a sensitive individual struggling to accept the unjust reality around him, and hoping to redress exploitation and racial inequality. Differing radically in the set of obstacles they try to negotiate, both Shorty and Toussaint respond to the lived experience of oppression, exploitation and marginalisation and attempt to reconstruct a sense of wholeness, that of a 'full' individual and of a 'full' citizen. Where they do diverge is in their opposing reactions to the experience of violence and the course of action taken: the boy's self-destructive pursuit is in stark contrast to Toussaint's transformative aspirations. This disparity has a clear didactic aim as it signals that personal change, regardless of one's past, is always possible and that meaningful, and curative, restoration of self cannot be founded on violence. As the narrative unfolds, the adventures and experiences of the two characters echo, complement or anticipate each other across the two narrative levels. This increasing proximity culminates in the last ('Then') section of the novel. In this chapter, Toussaint's spirit travels across time from his prison cell to the hospital ruins in Port-au-Prince to take up residence in Shorty's body. They are first the new *marasa*, to then become the even stronger *dosu/dosa*:

> His [Toussaint's] journey had ended; his exodus was over.
> He *had* returned.
> But to where? [...]
> He was trapped in an impenetrable ruin, with something heavy bearing down on his leg. [...]
> He was in a cave, and there was no way out [italics original].[93]

In these closing lines of the chapter, Toussaint metaphorically becomes Shorty's 'other half', replacing the void the boy experienced after his sister's death and that he had been trying, unsuccessfully, to counter with violence. The symbolic possession that leads to an awakening and corporeal union between Shorty and Toussaint has a composite potential as it collectively joins the two nation-defining moments, 1804 and 2010, endowing them both with a positively transformative potential. The scene caps the novel's formal interplay of the two narrative threads; from being parallel then echoing and interlacing narrative strands, they are now joined in this highly symbolic scene. On a thematic level, this narrative climax, which brings together the present and the past, seems to also herald the coming of a collective *dosou/dosa* – an even more powerful third element, an anticipated time of change for all.

The plaintive collective

Yet the capacity of the metaphor to evoke a third potentiality and a vision of a renewed collective is only partially realised and somewhat curtailed by the novel's emphasis on the individual transformation of the young narrator: the revolutionary rage against injustice and the roar of *goudougoudou* are translated as Shorty's voiced resolve to become a better person. Moreover, the novel's generic emphasis on the focalised and individualised experience, no longer contextualised within the power dynamics and liminality of urban space but rather conceptualised in terms of belonging to a family unit, overturns, and partially undermines, the text's earlier consideration of structural constraints, such as poverty and marginalisation. These factors had previously determined Shorty's decision to join the gang and remain unchanged, if not aggravated, in the aftermath of the earthquake. As the novel draws to a close, these are no longer the central constraints. The boy is eventually rescued by the relief team, and is now ready to start a new life, guided by a firm resolve to do good and to reconcile himself with his mother and their family's past.

Infused with the Toussaint's spirit, Shorty becomes again whole, reborn to the light, and this new force allows him to overcome the difficulties he will almost certainly face upon his return to Cité Soleil:

> And I have feelings and a soul in my chest, and I can talk and laugh and cry just like a real person, and I'm capable of doing good things. I've fucked up in the past, oh yes, I know I have, but, Manman, I'll try to make you proud.[94]

Freed from the physical and metaphorical darkness, Shorty's earlier hardly audible cry gives way to life and presence-affirming laughter. Emotionally and physically dezombified, through the symbolic joining with Toussaint's spirit, and rescued by the search teams, Shorty is once again capable of becoming a protagonist of his own life; he awakens as 'the hero [who was] always dormant' in the zombie.[95] No longer trying to assert his subjectivity through violence and crime, even if still claiming an urban vocabulary, Shorty is now driven by a desire to become a better person and to atone for the hurt he had previously inflicted on his family and those around him: 'I kept shooting and I made a goal, but I'm not gonna shoot no more.'[96] Framed in this way, the experience of being buried under the rubble has a clear pedagogical dimension; Shorty is ready to face new challenges, and now 'there's nothing that can get in and hurt [him]'.[97]

Shorty's triple rebirth – rescue from the rubble, turn away from violence and reconciliatory reunion with his mom – is a culmination of a double trajectory, one starkly resembling Woolley's path, with the closing paragraphs of the novel returning to the metaphors of darkness and light:

> Yes, I died, over and over.
> But now I've been reborn.
> Yes …

> Yes ...
> Yes ...
> I was in darkness, but now I am in light.[98]

With this image of light and triumphant rebirth the novel draws to a close, offering the reader a cathartic experience of a passage from the underworld which leads to a clear sense of completion, fulfilment and arguably a happy ending (at least for the young protagonist). Shorty, now infused with Toussaint's spirit and inspired by his example, returns to his mother in a loving embrace – a gesture which echoes his earlier coming together with Toussaint, filling the void after losing his twin sister. No longer condemned to obscurity and once again capable of forming an intimate parent-child bond, the two characters are replenished by each other's presence and 'words of fierceness, words of joy, love, love, love'[99] that they exchange. The hope-filled tone of the ending adds to the balance-restoring qualities of the scene. The triple 'yes' echoes the triple affirmation of love, recalling the third more powerful element, *dosu/dosa*. The changed 'we' is founded, then, on the acceptance of the departure of a loved one, Shorty's twin sister, and a reconciled re-turn: a turn towards those who, bruised and wounded, are left behind to build a new life. It is a 'we' that is 'greater than the sum of our individual parts, and part and parcel of a communal whole. $1+1=3$'.[100] Still, this new 'we' depends on the other one, to answer back and share these 'words of fierceness, words of joy, love, love, love'.[101]

Whereas *In Darkness*'s narrative of personal and even family transformation follows a discernible teleological, and possibly didactic, movement, its vision of collective recovery does not necessarily replicate these more pedagogical and hopeful designs. In keeping with its figurative use of spirituality and metaphorical transformation, Lake's work sees Haiti's revolutionary past and its religious tradition, which cannot be separated from the cultural memory of slavery and struggle for emancipation, as sources for this potential collective remaking. Yet to what extent does this pairing of 1804 and 2010, and of Toussaint and Shorty's joint dezombification, establish a positive collective vision beyond the family unit? Does the pairing suggest that after the darkness, first that of slavery and now of total post-disaster devastation, will come a moment of light – the (re) establishment of Haiti? Will the *ouverture* of 1804 return in the form of a parallel ascent from the post-earthquake rubble?

The novel's penultimate chapter, entitled 'Always', is key to a consideration of the implications Lake's work might have for the politics of the disaster and its collective aftermath. The section follows on from the experience of Shorty and Toussaint becoming one, analysed above. The narrative voice adopts the first-person plural 'we', and this shift seems to suggest that the focalisation of the two parallel narratives will be replaced in the remaining chapters of the novel by some form of collective narration:

> We are in darkness.
> We are always in the darkness [...]

Far beyond our walls, far beyond the bounds that hold us, there are people who want to help. There are always people who want to help, but they are too far away, and we are too silent. Though we have control of our own body, can animate our limbs to touch the boundaries of our reality, we are powerless to break through our reality, powerless to go out into the light, where the masters live.

We are a slave [...]

There is no future and no past.

We are in the darkness.

We are one.[102]

The incantatory tone of the passage creates an uncanny presence that lends it a sense of urgency with the short, repeated sentences imbuing the extract with a measured cadence. Similarly, the corporeal imagery, along with the repeated use of the possessive pronoun 'our', seems to create an evocative link between the bodily powerlessness of the collective 'we' and that of the maimed Haitian state. At the same time, the sudden change of narrative voice has a defamiliarising effect on the reader, while still allowing the narrative to offer empowering definitions of collective agency in the aftermath of the earthquake: the concluding 'we are one' echoes the motto from Haiti's 1807 coat of arms 'l'union fait la force' – unity makes strength. The country's revolutionary legacy is again invoked in this section, suggesting the possibility of a collective, not just a family, coming together in the wake of the disaster: together we are strong. Two are becoming one, heralding, here too, the coming of the powerful *dosou/dosa*.

However, this united and affirmative statement, recalling Shorty's opening invocation, quickly fades away. The overwhelmingly sombre silence is only briefly interrupted; the call receives no response, bounces back in the darkness and remains unanswered; 'there is no one to listen'.[103] The silencing of this plea is further emphasised by the very imagery of light and darkness as well as the return, in the subsequent and final section, to Shorty's focalised narration, employed earlier to depict transformations of subjectivity and agency. First, this visual contrast, voiced in terms of the master-slave dialectic, seems to evoke the language of colonial discourses on race, which traditionally oppose the 'darkness' of blackness and of servile exploitation with the light and power, at least in economic terms, of 'whiteness'. Not only does this echo earlier racist discourses of Haiti's 'backwardness', but it also overwrites, with a binary positioning that mirrors designated stages of socio-economic progress, Haiti's history of slavery, foreign intervention, and the resulting unequal distribution of wealth and power among the nation's white and black (then black and mulatto) populations. Second, the 'we' employed in the passage does not refer to a specific group, but rather to some unidentified mass living in the obscure margins who are 'powerless to break through [their] reality to the "light where the masters live"'.[104] Here, the masters' light is precisely what these masses, devoid of individual identity, should aim for. Still, their voices, and their hopes, are stifled.

Consequently, contrary to its earlier critiques and a more radical stance, the novel seems to suggest here that the only solution to escape this ghostly predicament is to accept external aid. Liberation, if it exists, will come from the outside. It is in the hands of all those 'people who want to help'[105] but who are too far away to hear and respond to Haitians' desperate cries for support: 'Far beyond our walls, far beyond the bounds that hold us, there are people who want to help. There are always people who want to help, but they are too far away, and we are too silent.'[106] Suggesting collective powerlessness and the inability to break through obscurity, 'our reality',[107] into the light, the passage reverses the positive scenarios sketched earlier for Shorty, implying that, despite the obvious obstacles, it is still up to those trapped in this physical and metaphorical darkness to 'make themselves be heard' and 'let themselves be helped'. In one 'peculiar turn of the tongue',[108] as Kaiama L. Glover demonstrates in her astute analysis of Haiti as '*ur*-example of the Afro-abject',[109] disaster becomes 'a state of being, as opposed to an event',[110] or, for that matter, a manifestation of compound, multi-scalar vulnerabilities.

In effect, the collective 'we' no longer represents a group of citizens but a desperate and anonymous mass; they are receivers rather than initiators of transformation, partly to blame for their own predicament.[111] Read symbolically as the spirits of those who died in the earthquake, the 'we' is trapped under the rubble. Their *ti bon anj* captured by the masters, is unable to leave the liminal state and to be set free on a journey to the ancestral land of Ginen, a cosmic Africa, where the spirits live and reign. Rather, the only 'present' state and envisaged future is that of darkness; even though Shorty-Toussaint has undergone a metaphorical transformation; indeed the boy is rescued from the darkness; collective rebirth and recovery from darkness seem almost impossible. Gone is the revolutionary promise of the earlier twinned narratives, and the imprisoned horde is relegated to the margins and fixed as the ultimate abject 'other', seemingly far away. The image of the zombie horde then, in Glover's gloss, 'instantiates the inevitable violence of the earth's most wretched, those who, because of this abject state, are evacuated of political worth and revolutionary intent',[112] and are best kept confined, at a distance. Although 1804 might have been an instance of enlightenment, an opening leading onto collective liberation, in 2010 there is no such vision of a hopeful coming together and awakening imaginable, at least any time soon.

The dissonance is clear and so is the labour lying ahead: individual rescue is distinct from recovery and healing, both personal and collective. Whereas rescue from darkness can indeed mark a new opening, *l'ouverture*, and a start of a 'saved' life for Shorty, for the collective the 'decay of trapped things'[113] is all there is: the experience of exploitation and imprisonment is unending and ruination seems timeless. Recovery and healing, if possible for the collective at all, have a distinct cadence, away from the measured and rhythmic march, in the romantic and heroic mode, towards a brighter future. Emplotted within such a defined trajectory of the romance genre, narratives of the past, and for David Scott specifically 'anticolonial stories about past, present and future',[114:

tended to be narratives of overcoming, often narratives of vindication; they have tended to enact a distinctive rhythm and pacing, a distinctive direction, and to tell stories of salvation and redemption. They have largely depended upon a certain (utopian) horizon toward which the emancipationist history is imagined to be moving.[115]

Such design, more hopeful and teleological, is discernible in the narrative of Shorty's rescue: reborn into light and full of resolve, he can now lead a new life. Yet, the collective does not share the count and tempo of this movement of resolute overcoming. In contrast to the 'story-potential' of the romance genre, their horizon and their prospects are limited to the confines of darkness. In a reverse mode to the redeeming story line of individual opening, here the narrative of the collective present draws an enclosure, bereft of hope.

The discord between the two moves, if at times bordering on visions of disempowering changelessness, makes clearly audible the difficulty of transforming, as in renewing and healing, the post-disaster present. Resisting the naïve suggestion of a quick fix available to all – if only the right person, a new hero, comes – the 'we' section, ultimately, struggles to find the tone and words for a collective future otherwise than as a halt and a stilled, mournful time. Only Shorty, who returns as the sole narrator of the book's last chapter, 'Now', is freed from this all-encompassing darkness. Following a change of heart, he is 'capable of doing good things',[116] has come to forgive his mother for concealing Marguerite's death, and can happily be reunited with her amidst expressions of mutual love. For the collective, however, 'there is no future and no past'.[117] The 'we', not yet 'counted as human',[118] remains 'in the darkness'[119] – no one answers their calls with laughter and words of love and joy – as the narrative draws to a plaintive close.

Diasporic *Marasa*

Whereas Lake's *In Darkness* suggests, at least, a partial possibility of post-earthquake redefinition of self through replacement and twinning with a 'new half', Wagner's *Hold Tight, Don't Let Go* (2015) contemplates individual and collective healing, precisely, in the absence and impossibility of being reunited with the other, *marasa*. The novel is composed of a series of only approximately dated, by month rather than day, diary entries of a young girl, Magdalie, and also begins on 12 January 2010. Yet it is more concerned than Lake's text, with the long and still unfolding aftermath of the disaster. *Hold Tight!* recounts the life of two cousins, Nadine and Magdalie, who survive the earthquake and are now trying to find again a post-disaster everyday among the debris and the reality of displacement and departure. Like Shorty and his sister, the cousins are, too, *marasa*, who complement each other, 'like two halves of a whole',[120] and 'are better and more powerful for being together'.[121] Their bond and complementarity, experienced at both physical and emotional levels, is the only thing that helps Magdalie face, and slowly come to terms with, the loss of her mom in the earthquake: 'the sound of [Nadine's] voice, the warmth and shape of her'[122] fill the void with which Magdalie is violently confronted. Yet Nadine's reassuring presence, and the little strength that it provides, are also cut short by the news of the cousin's departure to America; following a successful granting of the U.S. visa Nadine will join her father, who had left for the U.S. when she was eight years old. For Magdalie, her cousin's departure, 'her passage to lòt bò dlo, the other side of the water',[123] is a double blow: the move to America, across the water, is comparable, in some ways, to one's final journey, passing away, *anba dlo*, beneath the waters. The possibility of ever being a united whole again, a *marasa*, seems unconceivable.

Confronted with yet another loss, one that halves and shatters her sense of self, Magdalie no longer has the strength of *marasa*. Emptied, she is left on her own to find her way among the debris, an ongoing and slow process of *remnant dwelling*: that is, at once, a sense of living through the experience of violent fragmentation and an attempt to live through the fact of one's own survival and others' absence. This dual quotidian struggle is defined, in terms similar to that of Martin's memoir, by the experience of a stifling wait:

> We don't do much of anything but wait. Living in a camps is all about waiting – for food, for water, for something to change. [...] it's hard in a drudging, dragging way. We are fighting for our survival, but the fight is tedious and slow-motion.[124]

The only thing that breaks this disempowering inertia are, it seems, the varied forms of piercing everyday violence: Nadine's departure, cholera outbreak, Magdalie's unsuccessful attempts to earn a living, or her experience of being sexually assaulted. In response to these layered violences and in an attempt to somehow counter the hurt they leave behind, the protagonist, similarly to Shorty, turns to anger and silent rage, sentiments that, at least temporarily, fill the bruised void. For her, the earthquake 'broke open all the sadness in [her] heart' which she then took and turned 'into hate, because it made [her] feel strong instead of weak'.[125] In effect, these feelings allow Magdalie to gather the necessary strength that she desperately needs in order to keep going and endure the hurtful everyday.

At the same time, these sentiments, in the long-term, cannot alleviate the pain of Nadine's and Manman's absence. Rather than providing her with a sense of relief, as she slowly comes to realise, they force her to re-live and indwell her loss. Where for Shorty the symbolic twinning with Haiti's revolutionary leader allows him to translate violence into a resolve to lead a new life, for Magdalie such transition starts with an encounter with two Vodou *lwa*. First is Gede. The 'raucous Vodou spirit (or, more properly, family of spirits) whose concerns are life's beginnings and end – death and regeneration, ancestors and progeny'[126] – prepares Magdalie for this journey from death, to afterlife and to life again. The *lwa* is a 'grand master of transformation, embraces as life, as part of death, rebirth, and regeneration'.[127] Gede is, at once, the 'lord of the cemetery and the porter of the spirits of the dead to the outer realm';[128] he accompanies the dead on their last crossing. The second *lwa* the protagonist encounters is Erzulie Dantò, one of the two images of Ezili in the Vodou pantheon, a 'Black Madonna with a militant persona whose impetuousness defends her children violently against oppression'.[129] Seen, by Colin (Joan) Dayan, as a god specific to Haiti, whose strangest attributes delineate a history of women during slavery, Erzulie 'continues to embody a memory of slavery, intimacy, and revenge'.[130] She visits Magdalie in her dreams and then again mounts the girl during a ritual in *lakou* which, in its dual geographical and spiritual significance, 'includes the extended family's farm property, individual dwellings, and communal worship area'[131] and represents the intersection between land, the extended family and spirituality.[132] The two encounters with the *lwa* set off a slow process of confrontation with and acknowledgement of her double loss, enabling her, for the first time, to shed the long-repressed tears of grief and make her own journey, later on, from the depths of her enraged bereavement towards some, even partial, acceptance of the departure of the loved ones, *lot bò dlo*, the other side of the water.[133]

The Haitian Kreyòl phrase, *lot bò dlo*, evokes not only the geographical displacement, that of Nadine's family – as well as that of generations of Haitians often reduced to the news-headline tag 'boat people', escaping political persecution, violence and economic marginalisation – from the precarious location to 'the longed-for destination of promise and refuge, but also, more perilously, death'.[134] The crossing of waters, for '[us, Haitians,] who simply call ourselves/ People',[135] in its second meaning, also refers to an end of one's life journey, sinking beneath the waters, '*anba dlo*, that one is now fallen into a watery grave where ancestral spirits earlier went to rest during the Middle Passage'.[136] Departure, then, in its dual sense as journey across and beneath the waters from the unliveable present, is always marked by the histories of the millions of unrecovered bodies and lives of those who, as Christina Sharpe captures in a piercing prose, 'passed through the doors of no return[,] did not survive the holding and the sea [yet], they, like us, are alive in hydrogen, in oxygen; in carbon, in phosphorous, and iron; in sodium and chlorine'.[137] Moreover, the promises and gains of emigration, and the hopes for the other's return, are necessarily tainted by the baleful realisation that one's current displacement prefigures and foreshadows, in some ways, their permanent absence. For Magdalie, this encounter and experience of being mounted by the *lwa*, strengthened later by a community ritual of mourning and of setting her mother's soul free, marks the beginning of the protagonist's move from rage towards reconciliation. Still, this movement does not lead to a restoration of an at least partially healed whole, like in Lake's text, but, rather, relies on the acceptance of a double, most likely permanent, absence.

The girl is made to confront her loss and sense of a splintered self during community ceremonies mourning the death of her mother:

The last day of the nine-days long rituals marks this important threshold:
This is the last day we are allowed to cry for Manman. This is the end of mourning, and it marks the moment when we are supposed to move on. We have liberated her spirit from where it dwelled in the cold underwater.

We have released her from this mortal life. [...] Now I am sad because her death isn't new anymore, and because the world and my life have continued to turn without her. There was a comfort when the wound was fresh, because it meant that Manman was [p. 220] here – that she existed, that she was important, that she was part of our everyday life. Now we have not only buried her, but we've buried the loss itself.[138]

The ritual marking of one's life as accounted for, honoured and passed, accompanied by the releasing of Manman's spirit, creates a moment of double release for Magdalie from her anger and the concomitant sense of paralysing entrapment. At the same time, the protagonist is reluctant to let go with the verb 'supposed to' directly pointing to the dissonance between the temporalities of communal mourning and open-ended, seemingly endless, personal grief. This two-step act of letting go is only a start, rather than a culmination, of a long and laboursome process of creating one's dwelling among remnants of the painful recent past, the memories of all who are gone and the material reality of all the destruction the disaster left behind.

Paradoxically, the acceptance of Nadine's departure does not provide the same sense of a potentially, and eventually, restorative closure. Rather, the absent cousin, a liminal figure, is always painfully present in the protagonist's life; Nadine 'is present and not present. She is connected to [Magdalie's] life, but it is not the same thing at all'.[139] These dispersed and diasporic ties, in effect, acquire a dual character: although they too provide a sense of connection to her other half, they equally bind the protagonist to a longed-for yet unachievable past of shared lives, laughs and tears. Although the diary-form of the narrative necessarily privileges Magdalie's perspective, this twin experience of dispersal and alienation, of being and living '*dyaspora*',[140] is one also imposed on Nadine: she had no choice but to join her father in the United States. In this context, Magdalie's resentfulness towards her absent cousin is more an expression of their respective powerlessness in the face of involuntary separation. It is all the more painful since the girls' reunion is theoretically conceivable – the U.S. is not so far away and it is enough to get a visa and a ticket – yet practically unachievable. The required visa and the ticket are unattainable objects of desire and hope as well as symbols of power that represent the unequal and regulated mobilities of the girls' unrooted and de-routed lives. The country's border is a clear and violent reminder of the limits to one's dreams and aspirations, and the constraints to *dyasporic* healing, both for those who had left and those who are left behind. The fissure of post-earthquake departure is deepened by the pain of a conceivable yet highly unlikely reunion and the consequent sense of drifting apart and fading away of once shared lives. In effect, this double impossibility invalidates any notions of post-earthquake recovery of self as a return to an unshattered pre-disaster past: the permanent loss of her mom and then Nadine's departure can neither be undone nor forgotten.

Fully aware of this, Magdalie refuses, despite all, to give in to desperation of ruptured present and separated future. Her remnant dwelling is the remaking,

making anew, rather than recovering of an 'unbroken life'. It manifests itself in the everyday defiance and a struggle not to give in nor give up. Although life, for the protagonist, feels 'just like playing lotto against someone who rigged the game', she still keeps 'playing anyway – pooling [her] coins every week in hope of an eventual payout, even though [she knows] the system is against [her]'.[141] This resolve becomes an act and a process of resistance, one rooted in the diurnal struggle not divine intervention. It is a conscious stance in the world and a refusal to be silenced and immobilised in one's voiceless and disempowered position with the closing lines of her account full of this strong, yet realistic resolve: 'I don't need a miracle [...] I just need a chance to gather up my wishes, to write my own ending, in which everything is the way it's supposed to be.'[142]

Still, 'the writing of [one's] own ending' is an ongoing, open-ended process, one which does not necessarily bring resolution or closure nor complete healing of one's many wounds. The last entry of the book, 'January 2020', is Magdalie's dream of her 'impossible imagined future'.[143] In it, she can see Haiti fully cleared of the earthquake rubble and rebuilt, in receipt of reparations from France for the 1825 indemnity payment (meant to compensate France for the loss of its slave colony and slaves) which had been demanded by France in exchange for French recognition of the sovereign Haitian state following the Revolution. This last, dreamed, diary entry is also a moment of escape as well as a momentary reassurance that the arduous quotidian will, little by little, change. If nothing else, she hopes that, one day, she will be able to see Nadine again and share with her the stories of 'everything good that has ever happened in their lives'.[144] The happy ending imagined here, for the country and for herself, is ultimately a bittersweet one: for now, it only exits as a dream.

Conclusion

To conclude, the three texts draw, in varying degrees, on religious and spiritual symbolism, the ideas and aesthetics of twinning in their attempt to share the experience of the earthquake, communicate the joint sentiments of pain, grief and loss, and to envisage possible ways of formulating anew one's sense of self and place within the wider collective 'we'; whether that of a family unit, fellow survivors or a national community. Central to these considerations across the three, otherwise generically and aesthetically different, texts are notions of (im)possible recovery, restoration and return – ideas that also reveal surprising points of connection between them. One such shared ambition is the willingness to contribute in tangible, monetary, ways to Haiti's post-earthquake reconstruction. Whereas Woolley embeds this ambition within the logic of religiously-motivated fundraising as compassion, Lake makes a less personal appeal by pointing to charities worth supporting, in the accompanying material to the book. These two narratives of improvement draw the individual reader, in turn, into a journey of self-betterment that translates more or less explicitly into an appeal for financial support. Consequently, the narrative *dénouement* is not limited to the narrator's conversion, but also includes the reader's active response, both spiritual and financial, to the books' appeal. Wagner's novel, for its part, is already a fundraising object: a portion of the proceeds from the sale of the book directly go to the CES (Centre d'Education Spéciale; Special Education Centre) and Ti Kay Haiti organisation, in support of their work in education and health-care in Haiti. In effect, the texts themselves become sites of double twinning. First, as non-fiction and fictional narratives they join the reader's experience of his/her 'here' and 'now' to the 'there' and 'then' of the disaster. Second, they create a firm connection between the world of the text, whether a testimony or a fictional narrative, and the material reality of post-earthquake Port-au-Prince.

Furthermore, the texts' consideration and portrayal of recovery and restor-ation also reveals further resonances across these twinned narratives. In both *Unshaken* and *In Darkness* the movement from darkness to light and from fragmentation to wholeness defines the formal qualities of the texts, with extra-textual material supporting this narrative arc. Despite considerable differences, both texts fall short of envisaging empowered models of collective post-earthquake agency. As they draw to a close, *Unshaken* and *In Darkness* return to individual transformations of subjectivity, which are presented, for Woolley, within a religious narrative of iterative metanoia and, for Lake, in more teleolo-gical, didactic terms: Woolley renews his commitment to Christian faith, and Shorty resolves to lead a life free of violence and crime. Yet, neither of these individual recoveries is a fixed nor a permanent one. In *Unshaken*, doubt and unbelief tone down the author-narrator's expressions of gratitude and renewal. Similarly, in Lake's novel, the balance-restoring happy ending of Shorty's nar-rative is constrained by the persisting structures of vulnerability criticised throughout the text yet somewhat put to the side and suspended as Shorty's account draws to a close. Paradoxically though, the collective 'we' is trapped in darkness by these very same dynamics of exploitation and marginalisation that Shorty, infused with Toussaint L'Ouverture's spirit, seems to be able to over-come. Still, for now, the 'we' is arrested in a ghostly impasse. There is no eman-cipatory twinning possible; their unanswered call continues to echo in darkness.

Sharing in its recognition of the enduring constraints and limitations foreclos-ing post-earthquake individual and collective recovery, *Hold Tight!* refuses to give in to a vision of an unalterable predicament of ruined entrapment. Instead, the narrative pauses in the experience of diasporic separation and suspends its last word. For now, the ending is just a dream, yet still a dream; with the girls' hope of a reunion sustaining them through the everyday reality of separation. The long present of the interim, the in-between, is a work-in-progress. It is an open-ended and ongoing labour of *recovering*, beyond the binaries of restored fullness or permanent and fractured void. Still, it is not a silenced or muted time. Rather, its quotidian rhythm, the life-affirming pulse, is marked by the beat of each syllable in a simple, repeated way of saying goodbye in Haitian Kreyòl: *Kenbe fèm, pa lage* (Hold tight, don't let go).

Notes

1   Nick Lake is an award-winning British author, who grew up in Luxembourg, completed his undergraduate and graduate university education in England (Oxford), and currently lives in Oxfordshire and works in London as the Publishing Director at HarperCollins, one of the world's largest publishing companies. *In Darkness* (2012) was his first young adult novel, received critical acclaim and was the winner of the Michael L. Printz Award. Since its success, Lake has published the following novels: *The Blood Ninja* trilogy (2010–13); *Hostage Three* (2013) (which was named a *Publishers Weekly*, *School Library Journal* and *Boston Globe* Best Book of the Year); *There Will Be Lies* (2015) (A Boston Globe Best YA Book of 2015, A Texas TAYSHAS Pick); *Whisper To Me* (2016), a psychological thriller with a teenage heroine set in New Jersey. Lake's latest book, *Satellite,* was published in 2017. 'Nick Lake', *Unitedagents.co.uk*, www.unitedagents.co.uk/nick-lake [accessed 17 February 2018]. In an interview on the publisher's (Bloomsbury) page, Lake explains how a news story of a boy rescued from the rubble in Haiti was a major source of inspiration for him to write the book; see: 'In Darkness', *Bloomsbury.com* www.bloomsbury.com/uk/in-darkness-9781408819951/ [accessed 17 February 2018].

2   The book was the winner of the 2013 Michael L. Printz award for Excellence in Young Adult Literature and received a 2013 ALA Best Fiction for Young Readers listing.

3   *In Darkness* spells the Revolutionary leader's name as 'L'Ouverture' although the most commonly used spelling is Louverture. When discussing the book, however, I will keep the author's original spelling.

4   The following fragment of the letter is used for the epigraph: 'At the beginning of the troubles in Haiti, I felt that I was destined to do great things. When I received this divine intimation I was four-and-fifty years of age; I could neither read nor write.' Nick Lake, *In Darkness* (London: Bloomsbury, 2012). Epigraph. Although unacknowledged, this seemingly direct extract from L'Ouverture's letter to Bonaparte is actually a citation from John Relly Beard, *The Life of Toussaint L'Ouverture, The Negro Patriot of Hayti: Comprising an Account of the Struggle for Liberty in the Island, and a Sketch of Its History to the Present Period* (London: Ingram, Cooke, and Co., 1853), http://docsouth.unc.edu/neh/beardj/beard.html in University of North Carolina at Chapel Hill, *Documenting the American South: Primary Resources for the Study of Southern History, Literature, and Culture* http://docsouth.unc.edu/neh/beardj/beard.html [accessed 26 February 2015]. Beard, an English Unitarian minister, wrote the biography in order to 'supply the clearest evidence that there is no insuperable barrier between the light and the dark-coloured tribes of our common human species' (Beard, p. 1). In his work Beard argues for Louverture's supremacy as a military and political leader over George Washington or Bonaparte. He argues that Louverture is a better man than Bonaparte because 'the two differed in that which is the dividing line between the happy and the wretched; for while, with Bonaparte, God was a name, with Toussaint Louverture, God was at once the sole reality and the sovereign good' (Beard, p. 283). For Beard, Louverture's ultimate failure to liberate Haiti and his untimely death were the product of unfortunate circumstances – not an indictment of his character or leadership abilities. Zachary Hutchins, 'Summary', in *The Life of Toussaint L'Ouverture, the Negro Patriot of Hayti: Comprising an Account of the Struggle for Liberty in the Island, and a Sketch of Its History to the Present Period*, (Documenting the American South: Primary Resources for the Study of Southern History, Literature, and Culture: The University of North Carolina at Chapel Hill, 2004) http://docsouth.unc.edu/neh/beardj/summary.html [accessed 19 August 2015].

5   Laura Rose Wagner, *Hold Tight, Don't Let Go: A Novel of Haiti* (New York: Abrams, 2015), p. 18.

6 Wagner, book jacket.

7 Dinah Birch, 'young adult literature', in *The Oxford Companion to English Literature* (Oxford University Press, 2009) www.oxfordreference.com.wam.leeds. ac.uk/view/10.1093/acref/9780192806871.001.0001/acref-9780192806871-e-9563 [accessed 22 January 2015].

8 Jannice M. Alberghene, 'Will the Real Young Adult Novel Please Stand Up?', *Children's Literature Association Quarterly*, 10 (1985), 135–6 (p. 135).

9 Michael Cart, *From Romance to Realism: 50 Years of Growth and Change in Young Adult Literature* (New York, NY: HarperCollins, 1996), p. 8, in Amanda K. Allen, 'Breathlessly Awaiting the Next Installment: Revealing the Complexity of Young Adult Literature', *Children's Literature*, 40 (2012), 260–9 (p. 260).

10 Ruth Cline and William McBride, *A Guide to Literature for Young Adults: Background, Selection, and Use* (Glenview, IL: Scott, Foresman 1983), preface, in Alberghene, p. 135.

11 Allen, p. 260.

12 In their analysis of YAL, its merits beyond relevance to adolescents, and its perception by secondary school teachers and literary critics, Sotter and Connors provide an extensive list of young adult novels that engage with larger socio-political questions and conflicts:

> [...] we believe that young adult literature is capable of providing thoughtful social and political commentary that raises questions about complex issues – immigration (An Na's *A Step from Heaven*), the exploitation of children (Patricia McCormick's *Sold*), sexual orientation (M. E. Kerr's *Deliver Us from Evie*), terrorism (Cormier's *After the First Death*), roles of men and women in contemporary culture, social and political responsibility (M. T. Anderson's *Feed*), the individual challenge of social and political institutions (Suzanne Collin's *Hunger Games*), social conformity, religion (Pete Hatman's *Godless*), poverty, political morality (Walter Dean Myer's *Fallen Angels*), patriotism (Collier's *My Brother Sam Is Dead*), the strength of individuals to face disaster (Paul Volponi's *Hurricane Song: A Novel of New Orleans*) and the individual search of enduring truths (Gary Paulsen's *The Island*) among others.
>
> Anna O. Sotter and Sean P. Connors, 'Beyond Relevance to Literary Merit: Young Adult Literature as "Literature"', *The ALAN Review*, 37 (2009), 62–7 (pp. 63, 64)

13 Jeffrey S. Kaplan, 'Young Adult Literature in the 21st Century Moving Beyond Traditional Constraints and Conventions', *The ALAN Review*, 32 (2005), 11–18 (p. 11).

14 Stephen Roxburgh, 'The Art of the Young Adult Novel. Keynote Address: ALAN Workshop, Indianapolis, IN., November 20, 2004', *The ALAN Review*, 32 (2005), 4–10 (p. 5).

15 Roxburgh, p. 9.

16 Roxburgh, p. 8.

17 Roxburgh, p. 9.

18 In response to the 2010 earthquake two further publications, aimed specifically at children and young adults, were: Edwidge Danticat, *Eight Days: A Story of Haiti* (New York, NY: Orchard Books, 2010), and Ann E. Burg, *Serafina's Promise* (New York, NY: Scholastic Press, 2013).

19 Lake, p. 339.

20 Lake, supplement.

21 The following works (in this order) are listed on the page without any explanation or even a brief gloss on their content, scholarly value or relevance to the novel: *Thomas Clarkson: A Biography* by Ellen Gibson Wilson (William Sessions Ltd., 1996); *Rough Crossings: Britain, the Slaves and the American Revolution* by Simon Schama (BBC Books, 2006); *Bury the Chains: The British Struggle to Abolish Slavery* by Adam

Hochschild (Macmillan, 2005); *The Grand Slave Emporium: Cape Coast Castle and the British Slave Trade* by William St Clair (Profile Books, 2006); *Staying Power: The History of Black People in Britain* by Peter Fryer (Pluto Press, 1984); *Slaves Who Freed Haiti* by Katherine Scherman (1954); *The Black Jacobins* by C. L. R. James (1963) and *This Guilded African, Toussaint L'Ouverture* by Wenda Parkinson (1978).

22 Wagner, p. 257.

23 Laura Wagner, e-mails with the author, 3 February 2017.

24 *In Darkness.org* www.in-darkness.org/haiti [accessed 27 January 2015].

25 Wagner, book jacket.

26 Gurard Genette, *Narrative Discourse*, trans. by Jane E. Lewin and Jonathan Culler (Oxford: Basil Blackwell, 1980), p. 245.

27 In my analysis, I build upon Maria Nikolajeva's insights on the importance of approaching children's literature from the perspective of narrative theory, since it allows us to differentiate between the narrator and the focalisation of the text, and the reader's formation of subjectivity. See Maria Nikolajeva, 'Beyond the Grammar of Story, or How Can Children's Literature Criticism Benefit from Narrative Theory?', *Children's Literature Association Quarterly*, 28 (2003), 5–16 (p. 11).

28 Lake, p. 1.

29 Chris Baldick, 'narratee', in *The Oxford Dictionary of Literary Terms Online* (Oxford University Press, 2008) www.oxfordreference.com/10.1093/acref/9780199 208272.001.0001/acref-9780199208272-e-758 [accessed 29 January 2015]. The narratee is 'the fictive entity to which the narrator directs his narration' and can be further divided into two entities, namely the addressee and the recipient; whereas, 'the addressee is the narrator's image of the one to whom the message is sent, the recipient is the factual receiver'. Wolf Schmid, 'Narratee', in *The Living Handbook of Narratology Online*, ed. by Peter Hühn and others (Hamburg: Hamburg University, 2013) www.lhn.uni-hamburg.de/article/narratee [accessed 19 August 2015].

30 Lake, p. 16.

31 Maria Nikolajeva observes a general correspondence between the age of the narrator and his/her ability to express emotional states: 'a personal narrator who is a child telling the story more or less as it unfolds, in simultaneous personal narration, lacks both verbal and cognitive skills to articulate his emotions'. Nikolajeva, p. 12.

32 Lake, pp. 46–7.

33 In Lathan's gloss,

> [...] the magic represents an intrusion into an otherwise realistic environment, and paradoxically it is this merging of the magical and the real that serves to socialize the young adult reader by portraying an alternative – and perhaps subversive – view of society.
>
> Don Lathan, 'The Cultural Work of Magical Realism in Three Young Adult Novels', *Children's Literature in Education*, 38 (2007), 59–70 (p. 62)

34 Here is Shorty's invention:

> So I thought to myself, if there was a way to surround food with wet sand, maybe the water would dry from the sand and take the warmth from the food away with it. That was the story with the bowls [...]. – it was a fridge [...].
>
> Lake, p. 180

Thirteen pages later, the reader encounters a comparable description of Toussaint's creation: 'He had devised a simple system of two bowls, one inside the other, with watered sand between, which allowed food to remain cool for longer, and therefore not to spoil.' Lake, p. 193.

35 For the more attentive reader, the front cover of the book already foreshadows the narrative's engagement with Vodou, with vıvıs of Papa Legba and Erzulie visible against the cover's red background.

36 Philip Kaisary, in his study *The Haitian Revolution in the Literary Imaginary*, emphasises the role of the Revolution as an inspiration for a number of literary works, pieces of art and works of music from the nineteenth century onwards. Among them are literary works by William Wordsworth, Harriet Martineau, Victor Hugo, Alphonse de Lamartine, Henrich von Kleist, John Greenleaf Whittier; contemporary musical pieces such as Charles Mingus's 1957 *Haitian Fight Song*, Santana's 1971 *Toussaint L'Overture* (*sic*), Wyclef Jean's 2009 concept album *Toussaint St. Jean – From the Hut, to the Projects, to the Mansion*; and films such as Jean Negulesco's 1952 *Lydia Bailey* or *Toussaint* – a project by the American company Louverture films that has been long in development. Kaisary, pp. 4–8. Differentiating between 'radical' and 'conservative' works, Kaisary examines the ways in which 'certain aesthetic modes of recuperating the Haitian Revolution have enabled or hindered particular political visions'. Kaisary, p. 2.

37 Kaisary, p. 3.

38 Laënnec Hurbon, *Voodoo Search for the Spirit*, p. 53.

39 Hurbon, p. 56.

40 Laënnec Hurbon, 'American Fantasy and Haitian Vodou', in *Sacred Arts of Haitian Vodou*, ed. by Donald J. Cosentino (Los Angeles, CA: UCLA Fowler Museum of Cultural History, 1995), pp. 181–98 (p. 188).

41 Jon Stratton emphasises the rapid increase in the 2000s in the number of films and other media releases featuring zombies. Among the examples he gives are the video game *Resident Evil* (1996); Seth Grahame-Smith's retake on Austen's classic *Pride and Prejudice and Zombies* (Philadelphia, PA: Quirk Books, 2009); Steve Hockensmith's prequel *Pride and Prejudice and Zombies: Dawn of the Dreadfuls* (Philadelphia, PA: Quirk Books, 2010); and Danny Boyle's film *28 Days Later* (2002). Zombie-walk is a type of a public gathering, usually occurring in big cities, of people dressed up in zombie costumes. The idea originated in the United States and has since spread internationally. Some types of walks can have a charitable fundraising purpose, whereas others might be similar to a pub crawl or include a number of tasks and challenges set for the participants. There is now an international database of zombie-walks around the world and resources for those interested in organising a similar event. See: *Zombiewalk.com* http://zombiewalk.com/forum/index.php [accessed 14 May 2014].

42 For a comprehensive list of 'zombie films', see Shawn McIntosh and Marc Leverette, *Zombie Culture: Autopsies of the Living Dead* (Lanham, MD: Scarecrow Press, 2008), pp. 213–18.

43 For an analysis of the proliferation of zombies in popular North American contemporary culture, see Elizabeth McAlister, 'Slaves, Cannibals, and Infected Hyper-Whites: The Race and Religion of Zombies', *Anthropological Quarterly*, 85 (2012), 457–86.

44 The term 'zombie' denotes 'a company more dead than alive. Such a company is only kept alive by the goodwill and support of creditors, especially lenders'. Peter Moles and Nicholas Terry, 'zombie', in *The Handbook of International Financial Terms Online* (Oxford University Press, 1997) www.oxfordreference.com.wam. leeds.ac.uk/view/10.1093/acref/9780198294818.001.0001/acref-9780198294818-e-8501 [accessed 12 May 2014].

45 For Hurbon, the following early studies also contributed to the pseudo-scientific reductions of Vodou and zombification: Walter B. Cannon, ' "VOODOO" DEATH', *American Anthropologist*, 44 (1942); David Lester, 'Voodoo Death: Some New Thoughts on an Old Phenomenon', *American Anthropologist*, 74 (1972) 386–90; Harry D. Eastwell, 'Voodoo Death and the Mechanism for Dispatch of the Dying in East Arnhem, Australia', *American Anthropologist*, 84 (1982), 5–18. These, however, rather than engaging with Vodou as an Afro-Haitian system of thought, used 'voodoo death' as an umbrella term to analyse 'psycho-cultural', nonphysical causes of a death which 'characteristically occurs among native peoples after

putative sorcery, from taboo violation'. Eastwell, p. 5. As for Walter B. Cannon, he is treated as a key figure in physiological psychology, and his 1942 study is still hailed for its 'remarkable accuracy' 60 years later. See Esther M. Sternberg, 'Walter B. Cannon and "'Voodoo' Death": A Perspective From 60 Years On', *American Journal of Public Health*, 92 (2002), 1564–6.

46 These accusations have never been fully silenced. In one, more recent instance, President Donald Trump has repeated these very claims that 'all Haitians have AIDS' in the context of his decision to end Temporary Protected Status for Haitians, which 'allowed some 59,000 Haitians to live and work in the United States since an earthquake ravaged their country in 2010', and the U.S. government's wider anti-immigration programme. Miriam Jorden, 'Trump Administration Ends Temporary Protection for Haitians', *New York Times*, 20 November 2017 www.nytimes.com/2017/11/20/us/haitians-temporary-status.html [accessed 11 January 2018].

47 See, for example, Peter Moses and John Moses, 'Haiti and the Acquired Immunodeficiency Syndrome', *Annals of Internal Medicine*, 99 (1983), 565; William R. Greenfield, 'Night of the living dead II: Slow Virus Encephalopathies and AIDS: Do Necromantic Zombiists Transmit HTLV-III/LAV During Voodooistic Rituals?', *JAMA*, 256 (1986), 2199–200.

48 Paul Farmer, *AIDS and Accusation: Haiti and the Geography of Blame* (Berkeley and Los Angeles, CA: University of California Press, [1992], 2006), p. xii.

49 Elizabeth Abbott, *Haiti: The Duvaliers and Their Legacy* (New York, NY: McGraw-Hill, 1988), pp. 254–5.

50 Benjamin Hebblethwaite, 'The Scapegoating of Haitian Vodou Religion: David Brooks's (2010) Claim That "Voodoo" Is a "Progress-Resistant" Cultural Influence', *Journal of Black Studies*, 46 (2015), 3–22 (p. 6).

51 Harrison calls Vodou a 'progress-resistant force' and offers the following assessment of the religion's influence on Haiti:

> Voodoo [*sic*] is practiced mostly by poor Haitians, who make up the vast majority of the country's population. But all Haitians feel its influence, as one of my sons-in-law, who is Haitian and holds a graduate degree from Harvard, assures me. Wallace Hodges, an American missionary who lived in Haiti for 20 years, observed: 'A Haitian child is made to understand that everything that happens is due to the spirits. He is raised to externalize evil and to understand he is in continuous danger. Haitians are afraid of each other. You will find a high degree of paranoia in Haiti.' Brooks echoes Harrison's comments and adds lack of responsibility, bad child-rearing practices and inability to plan to the list: 'Haiti, like most of the world's poorest nations, suffers from a complex web of progress-resistant cultural influences. There is the influence of the voodoo religion, which spreads the message that life is capricious and planning futile. There are high levels of social mistrust. Responsibility is often not internalized. Child-rearing practices often involve neglect in the early years and harsh retribution when kids hit 9 or 10.
> See Brooks and Lawrence Harrison, 'Haiti and the Voodoo Curse: The Cultural Roots of the Country's Endless Misery', *Wall Street Journal*, 5 February 2010 http://online.wsj.com/news/articles/SB100014240527487045332045750471634 35348660 [accessed 25 February 2014]

52 Lawrence Harrison, *Underdevelopment is a State of Mind* (Lanham, MD: University Press of America, 1993), p. 105, in Paul Farmer, *The Uses of Haiti* (Monroe, ME: Common Courage Press, 2003), p. 285.

53 Somewhat more nuanced, such analytical frames still position Haiti's history as the main 'obstacle' to Haiti's progress and development, most often understood in economic terms: Mats Lundahl, 'History as an Obstacle to Change: The Case of Haiti', *Journal of Interamerican Studies and World Affairs*, 31 (1989), 1–21.

54 Hebblethwaite, p. 17.

55 Elizabeth McAlister, 'Slaves, Cannibals, and Infected Hyper-Whites', p. 473.

56 McAlister, p. 473.

57 Hurbon, p. 192.

58 Joan Dayan, *Haiti, History, and the Gods* (Berkeley, Los Angeles and London: University of California Press, 1995), p. 37.

59 Hurbon, p. 192.

60 Dayan, p. 37.

61 Kaiama L. Glover, *Haiti Unbound: A Spiralist Challenge to the Postcolonial Canon* (Liverpool: Liverpool University Press, 2010), p. 58.

62 Among some of the works that re-use the figure of the zombie from that period are Frankétienne, *Les affres d'un dfi* (Paris: Jean-Michel Place, 1979); Dany Laferrière, *Pays sans chapeau* (Montréal: Boréal, 1996); Gérard Etienne, *Le Nègre cruciféé* (Gèneve: Iditions Métropolis, 1974).

63 Indigenisme was an aesthetic and a political movement in Haiti. Its rise corresponded with the U.S. occupation of Haiti (1915–34). It 'inspired a renewed interest in and appreciation for Haiti's traditional culture', placed 'particular emphasis on the African roots of the peasantry's folk beliefs and practices as a valid source of creative inspiration', and 'encouraged a literary investment in the popular imagination – an imagination profoundly connected to the vodou faith'. Glover, p. 57. In his seminal *Ainsi Parla l'Oncle* (1928) (*So Spoke the Uncle*), Jean Price-Mars identified the causes of Haiti's

> national malaise as a failure to embrace its own traditions and a denial of the African origins of those traditions, such as the African roots of Vodou. He called for elevating the pride of Haitians and resisting US Occupation through the glorification of Haitian peasant traditions. This singular book is said to have given birth to the social and literary movement known as the Indigenist Movement.
>
> Garvey Lundy, 'Price-Mars, Jean', in *The Oxford Encyclopedia of African Thought*, ed. by Abiola Irele and Biodun Jeyifo (New York, NY: Oxford University Press, 2010), pp. 258–60 (pp. 259–60)

As a movement of opposition to the Occupation, Haitian indigeneism evolved around the 'poetics of resistance' (Munro, p. 12) and was predicated

> on the poetics of difference and on exploring an alternative genealogy to that imposed by Western culture. Haitian otherness became a fiercely nationalistic mask for confronting the nineteen-year occupation of Haiti by the U.S. Marines. [...] Indigenism was about beginning again, about creating a new methodology, a new history, where poets would be the keepers of knowledge. The rhetoric of power, therefore, reached well beyond the poetic act to envisage total, political solutions – or perhaps dissolutions – for a new, purified Haiti.
>
> Dash, p. 74; *Martin, Munro, Exile and Post-1946 Haitian Literature: Alexis, Depestre, Ollivier, Laferrière, Danticat* (Liverpool: Liverpool University Press, 2007), p. 12

64 Glover, p. 57.

65 Gerald L. Bruns, 'Becoming-Animal (Some Simple Ways)', *New Literary History*, 38 (2007), 703–20 (p. 704).

66 Glover, p. 59.

67 Glover, p. 59.

68 Here, it is important to recognise the importance of distinguishing between *zonbi* in Haitian Kreyòl and the bloodthirsty, cultural appropriation of it as zombie. Wherever possible, I use the Haitian Kreyòl spelling, not changing, however, the spelling in citations from Lake's book and secondary material. Among other critical works that discuss the use of the *zonbi*/zombie in the Haitian culture see, for example: Laënnec Hurbon, *Le barbare imaginaire* (*The Imaginary Barbarian*) (Paris: Cerf, 1988);

Kaiama L. Glover's, ' "Flesh like One's Own" '; Kate Ramsey *The Spirits and the Law*. Roger Luckhurst in his 2015 *Zombies: A Cultural History*, extensively discusses this ongoing phenomenon of 'zombie massification' (2015,b, p. 143) and the ways in which more recent uses of the zombie figure differ from 'their Haitian avatars as well as from the zombie's initial Hollywood iterations' (Glover, 'Flesh like One's Own', p. 253). Sarah Juliet Lauro, *The Transatlantic Zombie: Slavery, Rebellion, and Living Death* (New Brunswick: Rutgers University Press, 2015); Roger Luckhurst, *Zombies: A Cultural History* (London: Reaktion Book, 2015).

69 Lucy Swanson, 'Zombie Nation? The Horde, Social Uprisings, and National Narratives', *Cincinnati Romance Review*, 34 (2012), 13–33 (p. 14).

70 Hurbon, p. 192.

71 Hurbon, p. 192.

72 McAlister, p. 462.

73 Swanson, p. 15.

74 Glover, p. 58.

75 Kaiama L. Glover, ' "Flesh like One's Own": Benign Denials of Legitimate Complaint', *Public Culture*, 29 (2017), 235–60 (p. 251).

76 Colin (Jean) Dayan, *The Law is a White Dog: How Legal Rituals Make and Unmake Persons* (Princeton, NJ: Princeton University Press), p. 21.

77 Until 1975, Dahomey was the official name of the West African state now known as the Republic of Benin. The area equivalent to present-day Benin contained, during the precolonial period, several independent kingdoms. In the sixteenth and seventeenth centuries, the Kingdom of Allada was the dominant polity. A century later Dahomey developed into the most powerful kingdom. It was a French colony from the end of the nineteenth century and gained independence from France in 1960. The Fon people, most of whom practised the national religion Vodun, founded the kingdom of Dahomey. During the Atlantic slave trade, slaves shipped from Dahomey took the religion with them to the Americas. These traditional beliefs became an integral part of what is now the Haitian Vodou. See Tamba E. M'bayo, 'Dahomey', in *The Oxford Encyclopedia of African Thought*, pp. 269–71 (p. 269).

78 Toni Pressley-Sanon, 'One Plus One Equals Three: Marasa Consciousness, the Lwa, and Three Stories', *Research in African Literatures*, 44 (2013), 118–37 (p. 120).

79 Florence Bellande-Roberston, 'A Reading of the Marasa Concept in Lilas Desquiron's *Les Chemins de Loco-Miroir*', in *Haitian Vodou: Spirit, Myth, and Reality*, ed. by Patrick Bellegarde-Smith and Claudine Michel (Bloomington, IN: Indiana University Press 2006), pp. 103–11 (p. 104).

80 'In Voudoun one *and* one make three; two *and* two make five; for the *and* of the equation is the third and fifth part, respectively, the relationship which makes all the parts meaningful.' Maya Deren, *Divine Horsemen: The Living Gods of Haiti* (London and New York, NY: Thames and Hudson, 1953), p. 41.

81 Deren, p. 39.

82　[The] lwa are both related to and different from their West African progenitors. The religious systems of the Fon and the Yoruba, both of which made central contributions to Haitian Vodou, have complex pantheons of spirits. These spirits have hegemony over a wide variety of life domains, including natural phenomena such as thunder, wind, rain, and smallpox, as well as cultural activities such as farming and hunting. When these rich spiritual systems were transported to the Caribbean, their considerable power to make sense of the world came to focus almost exclusively on the most problematic arena of life there, the social arena.

Karen McCarthy Brown, 'Afro-Caribbean Spirituality: A Haitian Case Study', in *Vodou in Haitian Life and Culture: Invisible Powers*, ed. by Claudine Michel and Patrick Bellegarde-Smith (New York, NY: Palgrave Macmillan, 2006), pp. 1–27 (p. 17)

83 Bellande-Roberston, p. 105.
84 Bellande-Roberston, p. 105. As Leslie G. Desmangles explains: 'Legba is the divine medium through whom men's requests and prayers can be channeled to the respective *loas*. Legba is the interlocutor, the interpreter, the principle of crossing and of communication with the divine world.' Leslie G. Desmangles, 'African Interpretations of the Christian Cross in Vodou', in *Vodou in Haitian Life and Culture: Invisible Powers*, ed. by Claudine Michel and Patrick Bellegarde-Smith, pp. 39–50 (pp. 42–3).
85 Deren, p. 39. Also, I would like to thank Alessandra Benedicty-Kokken for pointing out to me the third twin's role to restore justice.
86 Shorty emphasises this sense of emptiness after his sister's disappearance throughout his account. See, for example: 'My sister, she was my twin. She was one half of me.' Lake, p. 8.

> [...] so when she was gone I became half a person. I would like you to remember this, so that you don't judge me later. Remember: even now, as I lie in this ruined hospital, I am only one half of a life, one half of a soul.
>
> Lake, p. 9

87 This is a highly complex term that means both 'a real ghost' and 'a real gangster'. In the contemporary Haitian context, the designation also has obvious political connotations as it was used in the mainstream media as a derogatory term for Jean-Bertrand Aristide's supporters, who were accused of using violence against Aristide opponents.
88 'Anyway, Route 9 came later. For now, all you need to know is that me and my family, we were living on a strop in the middle, between Dread Wilmè's territory and that of the rebels.

  You understand? No man's land. Manman told Papa we should leave, but he didn't want to.' Lake, p. 80.
89 Lake, p. 87.
90 The opening paragraphs of the novel establish this pairing between rescue and memory:

> I am the voice in the dark, calling out for your help.
>
> I am the quiet voice that you hope will not turn to silence, the voice you want to keep hearing cos it means someone is still alive. I am the voice calling for you to come and dig me out. I am the voice in the dark, asking you to unbury me, to bring me from the grave out into the light, like a zombi.
>
> I am a killer and I have been killed, too, over and over. [...]
>
> I have no name. There are no names in the darkness cos there is no one else, only me, and I already know who I am (I am the voice in the dark, calling out for your help), and I have no questions for myself and no need to call upon myself for anything, except to remember.
>
> Lake, p. 1

For Shorty forgetting someone is synonymous with destroying that person: 'Not only have I lost her [his mother] in the darkness, but I've also lost the memory of her. She's destroyed completely.' Lake, p. 329.
91 Lake, p. 8.
92 Lake, p. 93.
93 Lake, pp. 324–5.
94 Lake, p. 331.
95 Kaiama L. Glover, *Haiti Unbound*, p. 58.
96 Lake, p. 337.
97 Lake, p. 337.
98 Lake, p. 337.

99   Lake, p. 337.

100 Claudine Michel and Patrick Bellegarde-Smith, '*Koko Na Pa De Pwèl*: *Marasa* Reflection on Gede', in *In Extremis: Death and Life in 21st-Century Haitian Art* (Los Angeles: Fowler Museum at UCLA, 2012), pp. 13–16 (p. 15).

101 Lake, p. 337.

102 Lake, pp. 326–7.

103 Lake, p. 326.

104 Lake, p. 327.

105 Lake, p. 326.

106 Lake, p. 326.

107 Lake, p. 327.

108 Glover, ' "Flesh like One's Own" ', p. 239.

109 Glover, p. 236.

110 Glover, p. 242.

111 This contrasts with the novel's critical considerations of humanitarianism, its relationship with military intervention, and the ways in which, under the guise of assistance, such development initiatives often mask exploitative rather than empowering politics.

112 Glover, p. 255.

113 Lake, p. 327.

114 David Scott, *Conscripts of Modernity: The Tragedy of Colonial Enlightenment* (Durham and London: Duke University Press, 2004), p. 8.

115 Scott, pp. 8–9.

116 Lake, p. 331.

117 Lake, p. 327.

118 Glover, p. 255.

119 Lake, p. 327.

120 Wagner, *Hold Tight*, p. 19.

121 Wagner, p. 19.

122 Wagner, p. 8.

123 Wagner, p. 39.

124 Wagner, p. 15.

125 Wagner, p. 146.

126 Katherine Smith, 'Genealogies of Gede', in *In Extremis: Death and Life in 21st-Century Haitian Art*, ed.by Donald J. Cosentino (Los Angeles: Fowler Museum at UCLA, 2012), pp. 85–99 (p. 85).

127 Claudine Michel and Patrick Bellegarde-Smith, '*Koko Na Pa De Pwèl*', p. 14.

128 Myron M. Beasley, 'Vodou, Penises and Bones Ritual performances of death and eroticism in the cemetery and the junk yard of Port-au-Prince', *Performance Research*, 15 (2010), 41–7 (p. 43).

129 Leslie Desmangles, '*Our Lady of Class Struggle: The Cult of the Virgin Mary in Haiti* by Terry Rey (Trenton. NJ. and Asmara, Eritrea: Africa World Press, Inc., 1999) [book review]', *Transforming Anthropology*, 10 (2001), 44–5 (p. 44).

130 Colin (Joan) Dayan, 'Erzulie: A Women's History of Haiti', *Research in African Literatures*, 25 (1994), 5–31 (p. 11).

131 Haiti Lab, 'Religious life in the lakou', *Law & Housing in Haiti* https://sites.duke.edu/lawandhousinginhaiti/historical-background/lakou-model/religious-life-in-the-lakou/ [accessed 15 December 2017].

132 Haiti Lab, 'Religious life in the lakou'.

133 Wagner, p. 158.

134 Jana Evans Braziel, *Duvalier's Ghosts: Race, Diaspora, and US Imperialism in Haitian Literatures* (Gainesville: University Press of Florida, 2010), p. 67.

135 Felix Morisseau-Leroy, 'Boat People', *Pen.org*, 10 January 2013, https://pen.org/tourist-and-boat-people/ [accessed 28 November 2017].

136 Braziel, p. 67.
137 Christina Sharpe, *In the Wake: On Blackness and Being* (Durham and London: Duke University Press, 2016), p. 19.
138 Wagner, pp. 219–20.
139 Wagner, p. 234.
140 Edwidge Danticat, 'Introduction' in *The Butterfly's Way: Voices from the Haitian Dyaspora in the United States*, ed. by Edwidge Danticat (New York, NY: Soho, 2001), pp. ix–xvii (p. xiv).
141 Wagner, pp. 242–3.
142 Wagner, p. 243.
143 Laura Wagner, e-mails with the author, 3 February 2017.
144 Wagner, *Hold Tight*, p. 249.

Rasanblaj

A future reassembled

A polyphonic and carefully intertwined account of one family's life, told by several voices and spanning across time, Louis-Philippe Dalembert's *The Other Side of the Sea* (2014 [1998]) hopes to capture the longer histories of violent departures, crossings and ruptures that shape life in the Caribbean. The other side of the sea, also here, refers to those who have passed on or gone away, seeking new, hopefully better, lives. While taking into 'account all those deaths – the physical ones on land and at sea as well as the psychological deaths, the ones that precede and follow departures',[1] the text equally opens up to a consideration of what the future holds for those on the other side of the sea as well as those in the wake and carrying on with their lives on the island's shores. Among all the attendant uncertainties, one thing is sure: the sea remains. Yet this reassuring timelessness of the sea is in stark contrast with the equally rhythmic regularity of environmental hazards in the region and the disastrous, seemingly unpredictable and aberrant caesura they can become. For one of the main characters, the Grandma of the family, it is clear that:

> this region has a problem. One day, it's a storm that ushers the ocean and squalls of rain into your home, without asking your permission, uproots hundred-year-old mapou trees that, until then, had known how to resist the destructive power of men. Another day, a volcano spews its lava in your face. Then an earthquake comes like a thief at night and surprises you when you're sound asleep, shaking even a monumental fortress like an ordinary coconut tree.[2]

The 'region's problem', its hazardous history, captured here in an illustrative and dryly ironic description, is the ongoing multi-scalar vulnerability – a knot of processes, practices, history and politics, omissions and neglect – that amplified the force of geological tremors and continues to render hazardous environmental phenomena deadly and disastrous. These are at once quotidian, common place, occurrences and unique, if recursive, events for those directly affected and whose lives will never be the same. Whereas seasonal storms might still give some warning prompting often last-minute preparations, earthquakes 'come like a thief at night', thrusting all those within their reverberating reach into a new, cracked time. Such insistent recurrence demands attention, for the past and the present, as well as careful tending to what the future holds.

Varying in their motivations, complexity and levels of self-reflexivity, the otherwise very different narratives of the 2010 Haiti earthquake also tend to their personal losses, the lives of those on the other side of the sea, *lot bò dlo*, those who have survived and are still surviving the disaster. 'Tend', as used here in relation to these post-disaster accounts, mobilises the word's two meanings. First, as in 'to give attention, attend to, listen and apply oneself to the care of' and, second, in its use to denote 'of a ship at anchor: [t]o swing round with the turn of the tide or wind' (*OED*). In line with these layers, narratives of the disaster give attention, care, and equally turn with, circle round and sway to and fro the epicentre of the January 2010 tremors. This joint haunting sense of repetitiveness and unpredictability, of arbitrariness of survival – life-saving or life-ending minutes, meters and seconds – and the consequent obligation to witness and to testify that it implies is shared by the texts gathered here. Taken together, the writings analysed across the chapters register the attempt to create an image, however incomplete, of the indelible mark the catastrophe left upon all those who witnessed it, whether at first hand (Dany Laferrière, Rodney Saint-Éloi, Lionel-Édouard Martin, Laura Rose Wagner and Dan Woolley) or at different levels of remove (Sandra Marquez Stathis and Nick Lake). Each of the narratives captures and puts forward a highly particular perspective and interpretative frame, hoping to come to terms and share some of the understanding of what happened on 12 January 2010 and in the days, months and years following upon it. In so doing, they demonstrate the importance of attending to the formal variety of narratives that have emerged in dialogue with the past, present and future of the January disaster.

Their respective attempts to take stock of the overwhelming devastation, assemble and bring together whatever is left behind of the unbroken – 'faultless' and 'fault-free' for Martin – life are part of the dual effort to bridge across the rift created by the January disaster and to imagine a future otherwise. The seven accounts assembled here in non-hierarchical, inclusive networks of relations, so as to confront, complement and, at times, counter each other, allow the readers to explore the foundational, and now completely fractured and destabilised, categories of time, space and self – building blocks of disaster and of personal/collective experience that would be radically altered by the disaster and its aftermath. The book too, then, emphasises this sense of interconnections between a subjective sense of the earthquake, as explored through its various narrative treatments, and the wider historical, political and environmental contexts for this particular disaster and for disasters as a whole. Despite their varying standpoints, often existing in counterpoint, theirs is a shared attempt to reassemble and give expression to the collection of experiences, impressions and voices encountered in and among the debris. Accordingly, each of the chapters and the book as a whole embodies and is driven by the conceptual work and aesthetics of *rasanblaj*, that is the joining of cross-disciplinary approaches and interpretative frames in a shared, yet differently formulated, undertaking to understand the highly

particular histories of the 2010 earthquake, with narratives of the disaster experience as a starting point. In turn, these rooted perspectives allow for a broader
rethinking of what disaster is, how it is lived through and what practices and
processes of remaking, reconstruction and recovery it calls forth.

This aesthetics of reassembling and regrouping takes on different forms
across and within the seven narratives. From Laferrière's collection of short,
itemised entries and Martin's psalmic requiem memoir through to Saint-Éloi's
retrospective narrative of double recall of Haiti's past and a reimagining of its
future, these texts mobilise the narrative form in an attempt to amass whatever
remains. Equally, they hope to record the voices of those who remain, mourn
and grieve all the while seeking to remake and make anew their lives, day in,
day out. Yet even the strongly teleological, at times confidently optimistic, and
more linear narratives aiming at a sense of closure (Marquez Stathis, Woolley or
Lake and Wagner) point to the limitations of such endeavours, revealing the
impossibility of offering a panoptic or a resolved view of the disaster. In the
process, contrary to the first sense of narrative fixity and seamless linearity, these
texts expose a mosaic narrative form where distinct pieces are brought together
in a seemingly fixed decorative pattern, whether it is the evangelical theological
tradition as it is for Woolley or the thematic trajectory of growing up as explored
by Lake and Wagner. Still, the cracks and gaps between each of the fragments
remain visible. Equally the ruptures of the earthquake, on the personal, collective
and structural levels, punctuate any sense of a clear, linear and seamless progress
from the disaster event to a disaster-free future. Together, the texts analysed
across the four parts of this book manifest the open-ended and iterative nature of
personal/collective claims of recovery and remaking – whether configured as
conversion, moral improvement or growing up – as well as the palimpsestic
nature of material processes and practices of reconstruction and rebuilding.

Consequently, anchored in the highly particular and individualised experience, these narratives register, at times counteractive, attempts to circumscribe
and confine the January 2010 events by trying to locate their meaning within
longer personal, collective frames of reference and experience while, at the same
time, demonstrating the impossibility of separating these specific events from
their longer, regional, global and structural frames and histories. Similarly, these
countering, two-directional movements are traceable and shape the revisions –
as in going over again and seeing anew – of past and present in the light of the
event and the attempts to envisage a future beyond and otherwise than as a
rhythmic recurrence of disasters. In addressing this extended temporality and the
untimely experience of the earthquake Martin's and Saint-Éloi's aftermath
accounts challenge the event's finished character. Although the tremors may be
over, the disaster is still present in the form of recurring traumatic memories as
well as the threat that the island's geological faults will open again – a highly
likely yet equally unpredictable event. These untimely memoirs do the work of
memory, *la mémoire* in French, and of double recall as *remembering* and
reclaiming: in the process of creating and recording memories of the recent past
they call into question and hope to transform the ways in which Haiti's history

has been discursively framed. In this context, it becomes clear that the earthquake has to be placed in the context of the country's national history and discursive framings of it. *Goudougoudou* is neither a detached and unique occurrence, nor just another in a long sequence of Haitian 'natural' disasters or one, unending 'tropical apocalypse'. Rather, the scale and impact of the January 2010 tremors reveal the ways in which long-term processes of ecological and economic exploitation directly contributed to the scale of the devastation. Still, these socio-environmental conditions are first and foremost experienced on the micro-scale and manifest themselves in quotidian precarity, unequal mobility and differentiated access to essential resources such as wood or electricity.

This focus on quotidian depravations, accretive vulnerabilities and daily acts of resistance is central to Laferrière's itemised collection in which he pauses over what is left, wakes with fellow survivors, mourning the losses but also affirming a life, and future lives possible, in spite of all. Against the rapidity of rebuilding and the swiftness of progress-oriented reconstruction, *Tout bouge* takes the time to take stock of what happened and hopes to make time for at least a brief consideration of how to create this 'new city that will compel us to enter a new life'.[3] The narrator's feisty resolve is clearly audible in each pause between the consecutive fragments, reverberating through Laferrière's reflections on the whirlwind of events in which he is caught. A very different resolve animates Marquez Stathis's account, a narrative pledge of commitment to assuring a better future for Junior. Here, the narrator's personal journey of self-exploration, triggered by the news of the earthquake and the consecutive decision to return to Haiti and find her friend from years ago, doubles as an attempt to create and chart a trajectory of upward mobility for Junior. Less hopeful than Laferrière about the possibility of new urban futures, Marquez Stathis's account turns to Haiti's rural landscapes, tying its vision of the country's post-earthquake reconstruction to a joint protection and recovery of the bucolic countryside. Here, what is rebuilt is a nostalgic vision of a harmless nature with the 2010 disaster primarily seen as an urban event. Such binary positioning and the linking of the tremors to the city allow the narrator, nonetheless, to confine the emotional impact the witnessed destruction had on her and focus her efforts on helping Junior find a way out of the material and metaphorical debris. The analysis's joint emphasis on discursive and natural histories of the 2010 disaster calls for a reconsideration of the definition of post-disaster reconstruction which, in order to be empowering and ethical, must work towards transforming the pre-earthquake conditions of multi-scalar vulnerability and discursive and epistemic marginalisation that have held both Haiti and Haitians back.

The closing part of the book takes up these very questions of knowledge, marginalisation, empowerment and individual/collective renewal, anchoring them in considerations of individual rescue and collective recovery and healing. As the three focalised, generically diverse narratives by Woolley, Lake and Wagner explore also on the personal level, there is no simple return possible to an undamaged pre-earthquake state. In this sense rescue from debris or darkness is only a start of recovery – whether framed within Vodou symbolism and reference to Haiti's history, the evangelical journey of conversion or the joint process of mourning and growing up – which emerges as an open-ended process of reassembling whatever remains and of remaining; of still 'playing the lotto against someone who's rigged the game',[4] and persisting in doing so. Consequently, despite clear generic and conceptual differences, these three texts share in their attempt to think through the state of slow healing and *remnant dwelling*: that is, at once, a sense of living through the experience of violent fragmentation and an attempt to live through the fact of one's own survival and others' absence, whether permanent or not.

For Woolley, this state is clearly interpreted as a divine call to lead a better life and to be a witness to God's unshakeable love by sharing with the readers and the wider audience the narrator's experience of double rescue: for him, from the ruins of Hotel Montana and, for his whole family, from the six dark years of his wife's depression. Interpreted within a strong religious frame, this realisation of the two, retrospectively twinned, life-saving interventions marks, for Woolley, a turn to and a turning point on the life-long journey of metanoia. Lake's and Wagner's novels also focus on individual attempts to find anew one's sense of self in the radically changed reality while also trying to envisage possible post-earthquake futures for the collective. *In Darkness* turns to Vodou cosmologies and draws on the rich symbolism of the figures of *zonbi* and *marasa* twins to creatively consider the ways in which the legacy of the 1791–1804 Haitian Revolution, symbolised in the figure of Toussaint Louverture, can be a source for post-earthquake recovery. The climactic scene of the novel imagines the formation of a new, reawakened self – *dosu/dosa*, the stronger third child coming after the twins – epitomised in the coming together of Louverture and the rescued Shorty. Infused with the spirit of the legendary leader, the boy, in a surprisingly similar way to Woolley's transition from darkness to light, is now ready to start a new life, reunited with his mother and away from the revenge and void-driven violence that have guided his life before the earthquake. Still, the wider collective – the 'we' that briefly is given a united voice in the text – remains trapped in darkness. Their calls, ignored and unanswered, rebound and wane away.

For Magdalie and Nadine, the protagonists of Wagner's *Hold Tight!*, such a positive and restorative vision of a wounded yet reconstituted self and of a reunited family as a refuge, seems unavailable. Instead, for now, the post-earthquake present and long-term future are marked by loss and separation. For Magdalie, unlike for Shorty and Woolley, individual rescue from the earthquake – of being alive but not being spared the wounding trauma of disaster, displacement and loss – does not lead to a sense of newly-found wholeness and completion. Rather, the building of a new life for herself in Haiti, 'with all the possibilities for love, romance, friendship, and a future that exist for her there',[5] begins with a gradual acceptance of the two departures *lòt bo dlo*: her mother's crossing to Ginen, the land of ancestral spirits, and Nadine's migration. Neither of the two can be stopped nor be undone. The difficulty then, for Magdalie, is precisely to cease to dwell on and to let go of her anger at her loss yet still to indwell, to hold tight and not let go, of the here and now. Her diurnal efforts to remake and make anew her world constitute this open-ended work of slow healing. The cousins' respective struggles, their shared experience of separation and of being thrust into a new reality, of having to leave and of being left behind, is one shared by all so many other families in Haiti and those living in the *dyaspora*, all those 'exiles, émigrés, refugees, migrants, naturalised citizens, half-generation, first-generation'[6] who are trying, against all odds and all instances of quotidian and structural violence, to preserve and connect 'the worlds that have made [them]' [and] the connections and disconnections that have made [them] the mosaic that [they] are.[7]

As the preceding chapters have demonstrated, these distinct voices recorded, gathered, reconstructed and imagined across the seven texts come together in a collage of experiences, past memories, present wounds and dreams for the future. In their attempt to piece together the experience of the January 2010 tremors, these narratives consider questions essential to a broader and more nuanced understanding of disaster ontologies – including questions of what disasters are, how environmental phenomena become disasters and what types of relations they form – as well as their hinged chronologies. In other words, disasters emerge as a knot of structural histories, collective meaning-making frameworks and personal experience, one that is formed and unravels over time. Environmental events, such as the shifts of the tectonic plates, add to, make manifest and reveal pre-existing disastrous threads, whether conceptualised, in structural terms, under the category of vulnerability or lived through, on micro-scale, in the form of everyday deprivations. Yet these are not exclusive time-scales or realms of experience and are both part and different translations of the disaster experience, its processual, open-ended character, underpinned by multi-scalar vulnerabilities.

Therefore, the hinged chronologies of disaster experience necessitate and demand a conceptual as well as a practical frame that encompasses the long-term structural determinants of the disaster, its ongoing aftermath, and one that, in equal measure, tends to lives and futures lost and preserved. Such an approach, one that this book has hoped to offer, does not stop at interdisciplinarity but rather aims to be transformative across and within each of the interwoven disciplinary perspectives and their respective methodologies. This reassembled frame that this book calls for unsettles event-based, location-fixed conceptions of disaster and challenges teleological notions of reconstruction, recovery and healing. Such a frame builds on the analytical value of narrativised personal experience and emphasises the imprint of discursive constructs of Haiti on material practices and policies of aid and development. Equally, it argues for the importance of the imaginary – a category that encompasses both the fictional narratives and the hoped for visions of one's life as discernible in non-fictional accounts – for envisaging what is and what could be possible, differently.

Rooted in the narratives of the highly particular January 2010 disaster, this future-oriented paradigm is not limited, however, to Haiti. The conceptual and methodological insights offered here seek to forge a critical vocabulary, contributing to a 'living lexicon' of living through and living after the violent manifestations of disasters.[8] To this end, anchored in a series of attentive close-readings, the chapters have mobilised concepts of hinged chronologies, slow healing, unhealed futures and remnant dwelling. These encompass discussions of structural determinants of disasters and obstacles to recovery along with the consideration of the personal experience of long-term, wounding, disastrous histories and their manifestations, acknowledging the resulting asynchronous temporalities of a post-disaster present and future. These reassembled times cannot simply mean a return to the pre-disaster state, or hope for a naïve hazard-less future, but must entail the elimination of these vulnerabilities that created the disaster in the first place. Or, in the words of Kamau Brathwaite, 'it is not enough/to be pause, to be hole/to be void, to be silent/to be semicolon, to be semicolony'.[9] Guided by an open-ended vision and inspired by the resolve of the Haitian people, the conceptual and critical arguments presented here have been 'looking to possible futures, rather than to absolute endings'.[10] This scholarly labour hopes to 'draw others into a sense of curiosity and concern for our changing world',[11] opening up 'an appreciation of historical contingency: that things might have been and so might yet still be, otherwise'.[12] *Disasters, Vulnerability, and Narratives: Writing Haiti's Futures* is then one such contribution to this rigorously hopeful lexicon, one that can give 'words to refashion futures/like a healer's hand'.[13] In sum, an expression of such scholarship of care, this monograph works towards a vision of Haiti's and the Caribbean's futures otherwise than as a threat or fear.

Notes

1 Louis-Philippe Dalembert, *The Other Side of the Sea*, transl. by Robert H. McCormick Jr. (Charlottesville and London: University of Virginia Press, 2014), p. ix.

2 Dalembert, p. 16.

3 Homel, p. 158.

4 Wagner, *Hold Tight*, pp. 242–3.

5 Wagner, book jacket.

6 Danticat, 'Introduction' in *Butterfly's Way*, p. xv.

7 Joanne Hyppolite, 'Dyaspora' in *The Butterfly's Way*, pp. 7–12 (p. 11).

8 'Living Lexicon' is an initiative of the *Environmental Humanities* journal; see 'Living Lexicon', *Environmentalhumanities org* http://environmentalhumanities.org/lexicon/ [accessed 10 February 2018].

9 Kamau Brathwaite, 'VI Negus', *Islands* (London, New York, and Toronto: Oxford University Press, 1969), pp. 65–7, ll. 64–6.

10 Eben Kirksey, 'Hope', *Environmental Humanities*, 5 (2014), 295–300 (p. 295).

11 Thom van Dooren, 'Care', p. 294.

12 van Dooren, p. 293.

13 Kamau Brathwaite, 'VI Negus', ll. 64–6.

Bibliography

Abbott, Elizabeth, *Haiti: The Duvaliers and Their Legacy* (New York, NY: McGraw-Hill, 1988), pp. 254–5

Abbott, H. Porter, 'Narrative', in *The Encyclopedia of The Novel*, ed. by Peter Melville Logan (Oxford: Wiley Blackwell, 2011), pp. 533–40

'acropolis, n.', in *Oxford English Dictionary Online* (Oxford University Press, 2015) www.oed.com/view/Entry/1854?redirectedFrom=acropolis [accessed 17 January 2015]

Agamben, Giorgio, *State of Exception*, trans. by Kevin Attell (London and Chicago, IL: University of Chicago Press, 2005)

Agamben, Giorgio, *'The State of Exception'* (Extract from a lecture given at the Centre Roland-Barthes (Universite Paris VII, Denis-Diderot) and an edited translation of 'Lo stato di eccezione come paradigma di governo': the first chapter of Agamben's *Stato di eccezione. Homo Sacer II* (Bollati Boringhieri, May 2003, Torino www.egs.edu/faculty/giorgio-agamben/articles/state-of-exception/ [accessed 13 August 2015]

Alberghene, Jannice M., 'Will the Real Young Adult Novel Please Stand Up?', *Children's Literature Association Quarterly*, 10 (1985), 135–6

Alexander, David E., 'Resilience and disaster risk reduction: an etymological journey', *Natural Hazards and Earth System Sciences*, 13 (2013), 2707–16

Al Jazeera, 'Haitians Protest Home Demolition Plans', www.aljazeera.com/news/americas/2012/06/201262665643926283.html [accessed 09 September 2013]

Allen, Amanda K., 'Breathlessly Awaiting the Next Installment: Revealing the Complexity of Young Adult Literature', *Children's Literature*, 40 (2012), 260–9

Alpers, Paul, *What Is A Pastoral?* (Chicago, IL. and London: University of Chicago Press, 1996)

American Psychological Association, 'The road to resilience' (Washington, DC: American Psychological Association, 2014) www.apa.org/helpcenter/road-resilience.aspx [accessed 23 October 2017]

Anderson, Mark D., *Disaster Writing: The Cultural Politics of Catastrophe in Latin America* (Charlottesville and London: University of Virginia Press, 2011)

Aristide, Jean-Bertrand, *Dignity* (Charlottesville: University of Virginia Press, 1996)

Aristide, Jean-Bertrand, *Eyes of the Heart: Seeking a Path for the Poor in the Age of Globalization* (Monroe, ME: Common Courage Press, 2000)

Asante, Molefi Kete, 'Haiti: Three Analytical Narratives of Crisis and Recovery', *Journal of Black Studies*, 42 (2011), 76–87

Bailey, Sarah Pulliam, 'Pat Robertson: Haiti "Cursed" Since Pact with the Devil', *Christianity Today*, 13 January 2010 www.christianitytoday.com/gleanings/2010/january/

pat-robertson-haiti-cursed-since-pact-with-devil.html?paging=off [accessed 11 February 2015]

Bakun, William H., Claudia H. Flores, and Uri S. ten Brink, 'Significant Earthquakes on the Enriquillo Fault System, Hispaniola, 1500–2010: Implications for Seismic Hazard', *Bulletin of the Seismological Society of America*, 102 (2012), 18–30

Baldick, Chris, 'narratee', in *The Oxford Dictionary of Literary Terms Online* (Oxford University Press, 2008) www.oxfordreference.com/10.1093/acref/9780199208272.001.0001/acref-9780199208272-e-758 [accessed 29 January 2015]

Baldick, Chris, 'narrative', in *The Oxford Dictionary of Literary Terms Online* (Oxford University Press, 2008) www.oxfordreference.com.wam.leeds.ac.uk/view.10.1093/acref/9780199208272.001.0001/acref-9780199208272-e-760 [accessed 07 January 2014]

Ball, Nicole, 'The Myth of the Natural Disaster', *Ecologist*, 5 (1975), 368–9

Bankoff, Greg, *Cultures of Disaster: Society and Natural Hazard in the Philippines* (London and New York: Routledge, 2003)

Barthes Roland, *Camera Lucida: Reflections on Photography*, trans. by Richard Howard (New York, NY: Hill and Wang, 1981)

Beard, John Relly, *The Life of Toussaint L'Ouverture, The Negro Patriot of Hayti: Comprising an Account of the Struggle for Liberty in the Island, and a Sketch of Its History to the Present Period* (London: Ingram, Cooke, and Co., 1853) in University of North Carolina at Chapel Hill, *Documenting the American South: Primary Resources for the Study of Southern History, Literature, and Culture* http://docsouth.unc.edu/neh/beardj/beard.html [accessed 26 February 2015]

Beasley, Myron M., 'Vodou, Penises and Bones Ritual performances of death and eroticism in the cemetery and the junk yard of Port-au-Prince,' *Performance Research*, 15 (2010), 41–7

Beaudelaine Pierre and Nataša Ďurovičová, eds., *How to Write an Earthquake: Comment écrire et quoi écrire/Mo pou 12 janvye* (Iowa City, IA: Autumn Hill Books, 2011)

Bebbington, David, *Evangelicalism in Britain: a History from the 1730s to the 1980s* (London: Unwin Hyman, 1989)

Bell, Beverly, *Walking on Fire: Haitian Women's Stories of Survival and Resistance* (Ithaca, NY: Cornell University Press, 2001)

Bell, Beverly, 'The Business of Disaster: Where's the Haiti-Bound Money Going?', *Other Worlds Are Possible*, 8 April 2010 www.otherworldsarepossible.org/business-disaster-wheres-haiti-bound-money-going [accessed 07 January 2014]

Bell, Beverly, *Fault Lines: Views Across Haiti's Divide* (Ithaca, NY and London: Cornell University Press, 2013)

Bellande-Roberston, Florence, 'A Reading of the Marasa Concept in Lilas Desquiron's *Les Chemins de Loco-Miroir*', in *Haitian Vodou: Spirit, Myth, and Reality*, ed. by Patrick Bellegarde-Smith and Claudine Michel (Bloomington, IN: Indiana University Press, 2006), pp. 103–11

Bellegarde-Smith, Patrick, 'A Man-Made Disaster: The Earthquake of January 12, 2010-A Haitian Perspective', *Journal of Black Studies*, 42 (2011), 264–75

Bellegarde-Smith, Patrick and Claudine Michel, 'Introduction', in *Haitian Vodou: Spirit, Myth, and Reality* ed. by Patrick Bellegarde-Smith and Claudine Michel (Bloomington: Indiana University Press, 2006), xvii–xxvii

Benedicty, Alessandra, 'Aesthetics of "Ex-centricity" and Considerations of "Poverty"', *Small Axe*, 16 (2012), 166–76

Benedicty-Kokken, Alessandra, *Spirit Possession in French, Haitian, and Vodou Thought: An Intellectual History* (Lanham, MD: Lexington Books, 2015)

Benedicty-Kokken, Alessandra, Kaiama L. Glover, Mark Schuller and Jhon Picard Byron, eds., *The Haiti Exception: Anthropology and the Predicament of Narrative* (Liverpool: Liverpool University Press, 2016)

Benjamin, Walter, 'N: On the Theory of Knowledge, Theory of Progress', in *The Arcades Project*, ed. by Rolf Tiedemann, trans. by Howard Eiland and Kevin McLaughlin (Cambridge, MA. and London: The Belknap Press of Harvard University Press, 1999)

The Bible: New Revised Standard Version (Anglicised) (Oxford: Oxford University Press, 1995)

Birch, Dinah, 'young adult literature', in *The Oxford Companion to English Literature* (Oxford University Press, 2009) www.oxfordreference.com.wam.leeds.ac.uk/view/10.1093/acref/9780192806871.001.0001/acref-9780192806871-e-9563 [accessed 22 January 2015]

Blanc, Judite and others, 'Religious Beliefs, PTSD, Depression and Resilience in Survivors of the 2010 Haiti Earthquake', *Journal of Affective Disorders*, 190 (2016), 697–703

Blundell, D.J., 'Living with Earthquakes', *Disasters*, 1 (1977), 41–6

'body politic', in *New Oxford American Dictionary*, ed. by Angus Stevenson and Christine A. Lindberg (Oxford University Press, 2011) www.oxfordreference.com.wam.leeds.ac.uk/view/10.1093/acref/9780195392883.001.0001/m_en_us1227580?rskey=sNSZri&result=4 [accessed 29 August 2013]

Bonilla, Yarimar, *Non-Sovereign Futures: French Caribbean Politics in the Wake of Disenchantment* (Chicago and London: University of Chicago Press, 2015)

Bourguignon, Erika, 'Belief and Behavior in Haitian Folk Healing', in *Mental Health Services: The Cross-Cultural Context*, ed. by Paul B. Pedersen, Norman Sartorius, and Anthony J. Marsella (Beverly Hills, CA: Sage 1984), pp. 243–66

Bouley, Allan, 'Anamnesis', in *The New Dictionary of Theology*, ed. by Joseph, A. Komonchak, Mary Collins and Dermot A. Lane (Dublin: Gill and Macmillan, 1987), pp. 16–7

Bourgain, Anne, 'L'autre en soi ou la question du double chez un poète, '*Le Coq-héron*, 1 (2008), 97–104

Bowker, John, 'requiem', in *The Concise Oxford Dictionary of World Religions Online* (Oxford University Press, 2012) www.oxfordreference.com.wam.leeds.ac.uk/view/10.1093/acref/9780192800947.001.0001/acref-9780192800947-e-6057 [accessed 01 February 2013]

Brathwaite, Edward Kamau, *Islands* (London, New York, and Toronto: Oxford University Press, 1969)

Brathwaite, Edward Kamau, *History of the Voice: The Development of Nation Language in Anglophone Caribbean Poetry* (London: New Beacon, 1984)

Braziel, Jana Evans, *Duvalier's Ghosts: Race, Diaspora, and U.S. Imperialism in Haitian Literatures* (Gainesville: University Press of Florida, 2010)

Brereton, Virginia Lieson, *From Sin to Salvation: Stories of Women's Conversions, 1800 to the Present* (Bloomington, IN: Indiana University Press, 1991)

Breton, André, 'Préface', in *Derrière son double: Œuvres complètes*, ed. by François Di Dio (Paris: Gallimard, 1999), pp. 23–7

British Geological Survey, *Earthquakes* www.bgs.ac.uk/discoveringGeology/hazards/earthquakes/home.html [accessed 21 January 2014]

Brooks, David, 'The Underlying Tragedy', *New York Times*, 15 January 2010 www.nytimes.com/2010/01/15/opinion/15brooks.html?_r=0 [accessed 25 February 2014]

Bruns, Gerald L., 'Becoming-Animal (Some Simple Ways)', *New Literary History*, 38 (2007), 703–20

Buell, Lawrence, *The Environmental Imagination: Thoreau, Nature Writing, and the Formation of American Culture* (Cambridge, MA: Harvard University Press, 1995)

Cannon, Walter B., ' "VOODOO" DEATH', *American Anthropologist*, 44 (1942); David Lester, 'Voodoo Death: Some New Thoughts on an Old Phenomenon', *American Anthropologist*, 74 (1972) 386–90

Caroll, David, 'The Memory of Devastation and the Responsibilities of Thought: "And let's not talk about that" ', in *Heidegger and 'the Jews'*, trans. by Andreas Michel and Mark S. Roberts (Minneapolis, MN. and London: University of Minnesota Press, 1997), pp. vii–xxix

Carrigan, Anthony, *Postcolonial Tourism: Literature, Culture, and Environment* (New York: Routledge, 2011)

Carrigan, Anthony, 'Towards a Postcolonial Disaster Studies', in *Global Ecologies and the Environmental Humanities: Postcolonial Approaches*, ed. by Elizabeth DeLoughrey, Jill Didur, Anthony Carrigan (New York: Routledge, 2015), pp. 117–40

Cart, Michael, *From Romance to Realism: 50 Years of Growth and Change in Young Adult Literature* (New York, NY: HarperCollins, 1996)

'catastrophe', in *Oxford English Dictionary Online* (2012) www.oed.com/view/Entry/287 94?redirectedFrom=catastrophe#eid [accessed 12 December 2012]

Catechism of the Catholic Church (London: Chapman, 1994)

Center for Economic and Policy Research, *Haiti by the Numbers, Four Years Later* www. cepr.net/index.php/blogs/relief-and-reconstruction-watch/haiti-by-the-numbers-four-years-later [accessed 20 January 2014]

Césaire, Aimé, *Return to My Native Land*, trans. by John Berger and Anna Bostock (Baltimore: Penguin Books, 1969)

Césaire, Aimé, *Notebook of a Return to My Native Land/ Cahier d'un retour au pays natal*, trans. by Mireille Rosello with Annie Pritchard (Newcastle upon Tyne: Bloodaxe Books, 1995)

Césaire, Aimé, *Notebook of a Return to the Native Land*, trans. by Clayton Eshleman and Annette Smith (Middletown, CT: Wesleyan University Press, 2001)

Chambers, Ross, *Untimely Interventions: AIDS Writing, Testimonial, and the Rhetoric of Haunting* (Ann Arbor, MI: University of Michigan Press, 2004)

Chamoiseau Patrick, Raphaël Confiant, *Lettres créoles: Tracées antillaises et continentales de la littérature Haïti, Guadeloupe, Martinique, Guyane 1635*–1975 (Paris: Hatier, 1991)

'chronicle, n.', in *Oxford English Dictionary Online* (Oxford University Press, 2015) www. oed.com/view/Entry/32576?rskey=tBvYZE&result=1 [accessed 3 June 2015]

Cline, Ruth and William McBride, *A Guide to Literature for Young Adults: Background, Selection, and Use* (Glenview, IL: Scott, Foresman 1983)

Clitandre, Nadège T., 'Haitian Exceptionalism in the Caribbean and the Project of Rebuilding Haiti', *Journal of Haitian Studies*, 17 (2011), 146–53

Comay, Rebecca, 'Benjamin's Endgame', in *Walter Benjamin's Philosophy: Destruction and Experience*, ed. by Andrew Benjamin and Peter Osborne (London and New York, NY: Routledge, 1994), pp. 251–92

Compassion International, *About Us* www.compassion.com/about/about-us.htm [accessed 22 July 2015]

Compassion International, *History* www.compassion.com/history.htm [accessed 22 July 2015]

Compassion International, *Compassion International: Releasing Children from Poverty in Jesus' Name* www.compassion.com/sponsor_a_child/default.htm [accessed 9 January 2018] 'confessional', in *Merriam-Webster Dictionary Online* (2015) www.merriam-webster.com/dictionary/confessional [accessed 15 January 2015]

'confessional, adj.', in *Oxford English Dictionary Online* (Oxford University Press, 2014) www.oed.com/view/Entry/38782?rskey=mLXXcq&result=2&isAdvanced=false [accessed 13 November 2014]

'Confessions of St Augustine', in *The Concise Oxford Dictionary of the Christian Church Online*, ed. by Frank Leslie Cross and Elizabeth Anne Livingstone (Oxford University Press, 2005) www.oxfordreference.com/10.1093/oi/authority.20110803095631523 [accessed 3 March 2014]

Conway, Fred, 'Pentecostalism in the context of Haitian religion and health practice' (unpublished doctoral thesis, American University, 1978)

Corbett, Bob, *Haiti's Bad Press* http://haitibadpress.tumblr.com/ [accessed 25 February 2014]

Crapanzano, Vincent, *Waiting: The Whites of South Africa* (New York, NY: Vintage Books, 1986)

Crosby, Alfred W., *The Columbian Exchange: Biological and Cultural Consequences of 1492* (Westport, CT: Greenwood Press, 1972)

Dalembert, Louis-Philippe, *The Other Side of the Sea*, transl. by Robert H. McCormick Jr. (Charlottesville and London: University of Virginia Press, 2014)

Daniel, Trenton, 'Beauty vs Poverty: Haitian Slums Get Psychedelic Make-Over in Honour of Artist Prefete Duffaut's *Cities in the Sky*', *Independent*, 26 March 2013 www.independent.co.uk/arts-entertainment/art/news/beauty-vs-poverty-haitian-slums-get-psychedelic-makeover-in-honour-of-artist-prefete-duffauts-cities-in-the-sky-8549331.html [accessed 02 May 2013]

Danticat, Edwidge, 'Introduction' in *The Butterfly's Way: Voices from the Haitian Dyaspora in the United States* (New York, NY: Soho, 2001), pp. ix–xvii

Danticat, Edwidge, 'Lòt Bò Dlo, The Other Side of the Water', in *Haiti After the Earthquake*, ed. by Paul Farmer (New York, NY: Public Affairs, 2011), pp. 249–59

Das, Veena, Arthur Kleinman, Margaret M. Lock , Mamphela Ramphele and Pamela Reynolds, eds., *Remaking a World: Violence, Social Suffering and Recovery* (Berkeley: University of California Press, 2001)

Dash, J. Michael, *Haiti and the United States: National Stereotypes and Literary Imagination* (Basingstoke: Macmillan, 1988)

Dash, J. Michael, *The Other America: Caribbean Literature in a New World Context* (Charlottesville, VA. and London: University Press of Virginia, 1998)

Dayan, Colin (Jean), 'The Gods in the Trunk, or Writing in a Belittered World', *Yale French Studies*, 128 (2015), 92–112

Dayan, Colin (Jean), 'Erzulie: A Women's History of Haiti', *Research in African Literatures*, 25 (1994), 5–31

Dayan, Colin (Jean), *Haiti, History, and the Gods* (Berkeley, Los Angeles and London: University of California Press, 1995)

Dayan, Colin (Jean), *The Law is a White Dog: How Legal Rituals Make and Unmake Persons* (Princeton, NJ: Princeton University Press, 2011)

'délestrer', in *Reverso: Collins Online English–French Dictionary* (HarperCollins Publishers, 2012) http://dictionary.reverso.net/french-english/d%C3%A9lestrer [accessed 10 April 2013]

Delvecchio, Mary-Jo Good, Sandra Teresa Hyde, Sarah Pinto, and Byron J. Good, eds., *Postcolonial Disorders* (Berkeley: University of California Press, 2008)

Deren, Maya, *Divine Horsemen: The Living Gods of Haiti* (London and New York, NY: Thames and Hudson, 1953)

Desmangles, Leslie G., *The Faces of the Gods: Vodou and Roman Catholicism in Haiti* (Chapel Hill, NC: University of North Carolina Press, 1992)

Desmangles, Leslie G., 'Our Lady of Class Struggle: The Cult of the Virgin Mary in Haiti by Terry Rey (Trenton, NJ. and Asmara, Eritrea: Africa World Press, Inc., 1999) [book review]', *Transforming Anthropology*, 10 (2001), 44–5

Desmangles, Leslie G., 'African Interpretations of the Christian Cross in Vodou', in *Vodou in Haitian Life and Culture: Invisible Powers*, ed. by Claudine Michel and Patrick Bellegarde-Smith (New York, NY: Palgrave Macmillan, 2006), pp. 39–50

Díaz, Junot, 'Apocalypse: What Disasters Reveal', *Boston Review*, 1 May 2011, http://bostonreview.net/junot-diaz-apocalypse-haiti-earthquake [accessed 16 December 2017]

Disaster Tourism http://disastertourism.co.uk/disaster-tourism.html [accessed 13 March 2015]

Dorsey, Peter A., *Sacred Estrangement: The Rhetoric of Conversion in Modern American Autobiography* (University Park, PA: Pennsylvania State University Press, 1993)

Dorsinville, Nancy, 'Goudou Goudou', in *Haiti After the Earthquake*, ed. by Paul Farmer (New York, NY: Public Affairs, 2011), pp. 273–81

Douglas, Rachel, 'Writing the Haitian Earthquake and Creating Archives, *Continents manuscrits*, 8 (2017) http://journals.openedition.org/coma/859 [accessed 11 February 2018]

Dubois, Laurent, 'Vodou and History', *Comparative Studies in Society and History*, 43 (2001), 92–100

Duprey, Jean-Pierre, 'Trois feux et une tour', in *Derrière son double: Œuvres complètes*, ed. by François Di Dio (Paris: Gallimard, 1999), pp. 49–59

Duputel, Zacharie, Luis Rivera, Hiroo Kanamori, and Gavin P. Hayes, 'W phase source inversion for moderate to large earthquakes', *Geophysical Journal International*, 189 (2012), 1125–47

Dupuy, Alex, *The Prophet and Power: Jean-Bertrand Aristide, the International Community, and Haiti* (New York, NY: Rowman & Littlefield Publisher, 2007)

Duran, Isabel, 'Autobiography', in *The Literary Encyclopedia Online* www.litencyc.com/php/stopics.php?rec=true&UID=1232, [accessed 17 August 2015]

Duval, Frantz, 'Editorial', *Le Nouvelliste*, 29 September 2014 http://lenouvelliste.com/lenouvelliste/article/136410/Le-nouvel-ordre-geographique-et-administratif-dHaiti.html [accessed 17 November 2014]

Earthquake Track http://earthquaketrack.com/ [accessed 21 January 2014]

Eastwell, Harry D., 'Voodoo Death and the Mechanism for Dispatch of the Dying in East Arnhem, Australia', *American Anthropologist*, 84 (1982), 5–18

'émaner', in *Reverso: Collins Online English-French Dictionary* (HarperCollins Publishers, 2012) http://dictionary.reverso.net/french-english/émaner [accessed 22 November 2012]

Etienne, Harley F., 'Urban Planning and the Rebuilding of Port-au-Prince', in *The Idea of Haiti: Rethinking Crisis and Development*, ed. by Millery Polyné (Minneapolis and London: University of Minnesota Press, 2013), pp. 165–81

Fan, Lilianne, *Disaster as Opportunity? Building Back Better in Aceh, Myanmar and Haiti* (London: Humanitarian Policy Group, Overseas Development Institute, November 2013) www.odi.org.uk/publications/8007-resilience-build-back-better-bbb-aceh-tsunami-cyclone-myanmar-earthquake-haiti-disaster-recovery#downloads [accessed 13 February 2014]

Farmer, Paul, *The Uses of Haiti* (Monroe, ME: Common Courage Press, 2003)

Farmer, Paul, *AIDS and Accusation: Haiti and the Geography of Blame* (Berkeley and Los Angeles, CA: University of California Press, [1992], 2006)

Farmer, Paul, ed., *Haiti After the Earthquake*, (New York, NY: Public Affairs, 2011)

Fatton Jr. Robert, *Haiti's Predatory Republic: The Unending Transition to Democracy* (Boulder, CO: Lynne Rienner Publishers, 2002)

'Favela Painting Foundation', *Favelapainting.com*, www.favelapainting.com/ [accessed 16 November 2016]

'Favela Painting Investigation in Haiti', *Akvo.org* https://rsr.akvo.org/en/project/1855/#report [accessed 20 December 2016]

Fellmeth, Aaron X. and Maurice Horwitz, 'restitutio in integrum', in *Guide to Latin in International Law Online* (Oxford University Press, 2009) www.oxfordreference.com/ view/10.1093/acref/9780195369380.001.0001/acref-9780195369380-e-1856 [accessed 24 February 2015]

Ferris, Elizabeth and Sara Ferro-Ribeiro, 'Protecting People in Cities: the Disturbing Case of Haiti', *Disasters*, 36 (2012), 44–63

Figueroa, Víctor M., *Prophetic Visions of the Past: Pan-Caribbean Representations of the Haitian Revolution* (Columbus, OH: Ohio State University Press, 2015)

Fischer, Sibylle, 'Haiti: Fantasies of Bare Life', *Small Axe*, 11 (2007), 1–15

Fischer, Sibylle, 'Beyond Comprehension', *Social Text: Ayiti Kraze/Haiti in Fragments*, 26 January 2010, http://socialtextjournal.org/periscope_article/beyond_comprehension/ [accessed 1 August 2014]

Fludernik, Monika, *An Introduction to Narratology*, trans. by Patricia Häusler-Greenfield and Monika Fludernik (London and New York, NY: Routledge, 2009)

Folke, Carl, Stephen R. Carpenter, Brian Walker, Marten Scheffer, Terry Chapin and Johan Rockström, 'Resilience Thinking: Integrating Resilience, Adaptability and Transformability', *Ecology and Society*, 15 (2010) www.ecologyandsociety.org/vol. 15/iss4/art20/ [accessed 16 September 2017]

'French-Canadian Writers: Dany Laferrière', *Athabasca University*, http://canadian-writers.athabascau.ca/french/writers/dlaferriere/dlaferriere.php [accessed 10 February 2018]

Frenzel, Fabian, 'Slum tourism in the context of the tourism and poverty (relief) debate', *Die Erde: Journal of the Geographical Society of Berlin*, 144 (2013), 117–28

'Frequently Asked Question – Haiti Earthquake – 2010' (St. Augustine, Trinidad and Tobago: UWI, Seismic Research Centre) http://uwiseismic.com/Downloads/2010_01_HaitiEQ_FAQ2c.pdf [accessed 12 October 2017]

Füssel, Hans-Martin and Richard J. T. Klein, 'Climate Change Vulnerability Assessments: An Evolution of Conceptual Thinking', *Climatic Change*, 75 (2006) 301–29

Gagnon, Marie-Julie, '«Jalousie en couleur»: transformation extrême d'un bidonville', *Le Huffington Post Québec*, 16 May 2013 http://quebec.huffingtonpost.ca/2013/05/16/ jalousie-haiti_n_3287761.html [accessed 19 August 2013]

Gallagher, Catherine, 'Time', in *The Encyclopedia of The Novel*, ed. by Peter Melville Logan (Oxford: Wiley Blackwell, 2011), pp. 811–17

Gana, Nouri, 'Remembering Forbidding Mourning: Repetition, Indifference, Melanxiety, Hamlet', *Mosaic: A Journal For the Interdisciplinary Study of Literature*, 37 (2004), 59–78

Geggus, David Patrick and Norman Fiering, eds., *The World of the Haitian Revolution* (Bloomington, IN: Indiana University Press, 2009)

Gélin-Adams, Marlye and David M. Malone, 'Haiti: A Case of Endemic Weakness', in *State Failure and State Weakness in a Time of Terror*, ed. by Robert I. Rotberg (Cambridge, MA: World Peace Foundation; Washington, D.C: Brookings Institution Press, 2003), pp. 287–304

Gérard Genette, *Narrative Discourse*, trans. by Jane E. Lewin and Jonathan Culler (Oxford: Basil Blackwell, 1980)

Gifford, Terry, *Pastoral* (London and New York, NY: Routledge, 1999)

Gilmore, Leigh, *The Limits of Autobiography: Trauma and Testimony* (Ithaca, NY. and London: Cornell University Press, 2001)

Glantz, Michael H., 'Nine Fallacies of Natural Disaster: The Case of Sahel', *Climatic Change*, 1 (1977) 69–84

Glantz, Michael H., ed., *Creeping Environmental Problems and Sustainable Development In the Aral Sea Basin* (Cambridge: Cambridge University Press, 1999)

Glantz, Michael H. and Dale Jamieson, 'Societal response to Hurricane Mitch and intra- versus intergenerational equity issues: whose norms should apply?', *Risk Analysis*, 20 (2000), 869–82

Glover, Kaiama L., *Haiti Unbound: A Spiralist Challenge to the Postcolonial Canon* (Liverpool: Liverpool University Press, 2010)

Glover, Kaiama L., 'New Narratives of Haiti; or, How to Empathize with a Zombie', *Small Axe*, 16 (2012), 199–207

Glover, Kaiama L., '"Flesh like One's Own": Benign Denials of Legitimate Complaint', *Public Culture*, 29 (2017), 235–60

Gordillo, Gastón R., *Rubble; The Afterlife of Destruction* (Durham and London: Duke University Press, 2014)

Graziano, Frank, *Undocumented Dominican Migration* (Austin: University of Texas Press, 2013)

Greenfield, William R., 'Night of the living dead II: Slow Virus Encephalopathies and AIDS: Do Necromantic Zombiists Transmit HTLV-III/LAV During Voodooistic Rituals?', *JAMA*, 256 (1986), 2199–200

Griffin, Charles J. G., 'The rhetoric of form in conversion narratives', *Quarterly Journal of Speech*, 76 (1990), 152–63

Gros, Jean-Germain, 'Anatomy of a Haitian Tragedy: When the Fury of Nature Meets the Debility of the State', *Journal of Black Studies*, 42 (2011), 131–57

Groth, Helen and Paul Sheehan, 'Introduction: Timeliness and Untimeliness', *Textual Practice*, 26 (2012), 571–85

Guerda, Nicolas, Billie Schwartz and Elizabeth Pierre, 'Weathering the Storms Like Bamboo: The Strengths of Haitians in Coping with Natural Disasters', in *Mass Trauma and Emotional Healing Around the World: Rituals and Practices for Resilience and Meaning-Making*, ed. by Ani Kalayjian and Dominique Eugene (Santa Barbara, CA: Praeger, 2010), pp. 93–106

'Haiti Earthquake Facts and Figures', *Dec.org*, www.dec.org.uk/articles/haiti-earthquake-facts-and-figures [accessed 6 September 2017]

'Haiti Earthquake PDNA: Assessment of damage, losses, general and sectoral needs' (2010) https://siteresources.worldbank.org/INTLAC/Resources/PDNA_Haiti-2010_Working_Document_EN.pdf, p. 24 [accessed 12 October 2017]

Haiti Grassroots Watch, 'Martelly Government Betting on Sweatshops: Haiti: "Open for Business"', *Haïti Liberté*, 29 November 2011 www.haiti-liberte.com/archives/volume 5–20/Martelly%20government.asp [accessed 19 August 2013]

Haiti Grassroots Watch, 'Haiti "Open for Business": Sourcing Slave Labor for US-Based Companies: Is the Caracol Industrial Park Worth the Risk?', *Haïti Liberté*, 13 March

2013 www.globalresearch.ca/haiti-open-for-business-sourcing-slave-labor-for-u-s-based-companies/5327292 [accessed 19 August 2013]

'Haiti's History of Earthquakes', *Repeatingislands.com* https://repeatingislands.com/2010/01/15/haiti%E2%80%99s-history-of-earthquakes/ [accessed 21 February 2018]

Haiti Lab, 'Religious life in the lakou', *Law & Housing in Haiti* https://sites.duke.edu/lawandhousinginhaiti/historical-background/lakou-model/religious-life-in-the-lakou/ [accessed 15 December 2017]

Haiti Memory Project http://haitimemoryproject.org/ [accessed 14 May 2015]

'Haïti – Reconstruction: Deuxième phase des travaux de rénovation du quartier de Jalousie', *Haiti Libre*, 15 August 2013 www.haitilibre.com/article-9228-haiti-reconstruction-deuxieme-phase-des-travaux-de-renovation-du-quartier-de-jalousie.html [accessed 20 September 2013]

Haiti, *The World Factbook* www.cia.gov/library/publications/the-world-factbook/geos/ha.html [accessed 13 September 2017]

Hallward, Peter, *Damming the Flood: Haiti and the Politics of Containment* (London, New York, NY: Verso, 2010)

Harrison, Lawrence, *Underdevelopment is a State of Mind* (Lanham, MD: University Press of America, 1993)

Harrison, Lawrence, 'Haiti and the Voodoo Curse: The Cultural Roots of the Country's Endless Misery', *Wall Street Journal*, 5 February 2010 http://online.wsj.com/news/articles/SB10001424052748704533204575047163435348660 [accessed 25 February 2014]

Hayes, Gavin. P., Richard W. Briggs, Anthony Sladen, Eric J. Fielding, Carol Prentice, Kenneth Hudnut, Paul Mann, Frederick W. Taylor, A.J. Crone, Ryan Gold, Takeo Ito, and M. Simons, 'Complex Rupture During the 12 January 2010 Haiti Earthquake', *Nature Geoscience*, 3 (2010), 800–5

Hebblethwaite, Benjamin, 'The Scapegoating of Haitian Vodou Religion: David Brooks's (2010) Claim That "Voodoo" Is a "Progress-Resistant" Cultural Influence', *Journal of Black Studies*, 46 (2015), 3–22

Hewitt, Kenneth, 'The idea of calamity in a technocratic age', in *Interpretations of Calamity from the Viewpoint of Human Ecology*, ed. by Kenneth Hewitt (Boston, London, Sydney: Allen & Unwin Inc: 1983), pp. 3–32

Hewitt, Kenneth, *Regions of Risk: A Geographical Introduction to Disaster* (Harlow: Longman, 1997)

Hirsch, Marianne, 'The Generation of Postmemory', *Poetics Today*, 29 (2008), 102–28

Höges, Clemens, 'Preparing for the Next Earthquake: Haiti Debates Moving Its Capital', *Spiegel*, www.spiegel.de/international/world/preparing-for-the-next-earthquake-haiti-debates-moving-its-capital-a-675299-3.html, 2 February 2010 [accessed 14 December 2017]

Holling, Crawford Stanley, 'Resilience and Stability of Ecological Systems', *Annual Review of Ecology and Systematics*, 4 (1973), 1–23

Holm, Isak Winkel, 'Earthquake in Haiti: Kleist and the Birth of Modern Disaster Discourse', *New German Critique*, 39 (2012), 49–66

Huggan, Graham, ed., *Extreme Pursuits: Travel/Writing in an Age of Globalization* (Ann Arbor, MI: University of Michigan Press, 2009)

Huggan, Graham, ed., *The Oxford Handbook of Postcolonial Studies* (Oxford: Oxford University Press, 2013)

Huggan, Graham and Helen Tiffin, *Postcolonial Ecocriticism: Literature, Animals, Environment* (London and New York, NY: Routledge, 2010)

Human Rights Watch, *2001 Report*, www.hrw.org/legacy/wr2k1/americas/haiti.html [accessed 19 January 2016]

Hurbon, Laënnec, 'American Fantasy and Haitian Vodou', in *Sacred Arts of Haitian Vodou*, ed. by Donald J. Cosentino (Los Angeles, CA: UCLA Fowler Museum of Cultural History, 1995), pp. 181–98

Hurbon, Laënnec, *Voodoo Search for the Spirit* (New York, NY. and London: Harry N. Abrams, 1995)

Hutchins, Zachary, 'Summary', in *The Life of Toussaint L'Ouverture, the Negro Patriot of Hayti: Comprising an Account of the Struggle for Liberty in the Island, and a Sketch of Its History to the Present Period*, (Documenting the American South: Primary Resources for the Study of Southern History, Literature, and Culture: The University of North Carolina at Chapel Hill, 2004) http://docsouth.unc.edu/neh/beardj/summary. html [accessed 19 August 2015]

Hyppolite, Joanne, 'Dyaspora' in *The Butterfly's Way: Voices from the Haitian Dyaspora in the United States*, ed. by Edwidge Danticat (New York, NY: Soho, 2001), pp. 7–12

IJDH, 'Cholera Accountability', *IJDH.org*, www.ijdh.org/advocacies/our-work/cholera-advocacy/ [accessed 16 February 2018]

IJDH, 'Cholera Resources', *IJDH.org*, www.ijdh.org/cholera/cholera-resources/ [accessed 16 February 2018]

'In Darkness', *Bloomsbury.com* www.bloomsbury.com/uk/in-darkness-9781408819951/ [accessed 17 February 2018] *In Darkness.org* www.in-darkness.org/haiti [accessed 27 January 2015]

Ingram Jane C., Guillermo Franco, Cristina Rumbaitis-del Rio, and Bjian Khazai, 'Post-Disaster Recovery Dilemmas: Challenges in Balancing Short-Term and Long-Term Needs for Vulnerability Reduction', *Environmental Science & Policy*, 9 (2006), 607–13

L'Institut Haïtien de Statistique et d'Informatique, *Recensement Général de la population et de l'Habitat: Enquête nationale sur la population sa structure et ses caractéristiques démographiques et socio-économiques* (L'Institut Haïtien de Statistique et d'Informatique 2007) www.ihsi.ht/rgph_resultat_ensemble_population.htm [accessed 17 February 2015]

L'Institut Haïtien de Statistique et d'Informatique, 'Statistiques Démographiques & Sociales', www.ihsi.ht/accueil_presentation_general.htm [accessed 16 February 2018]

James, Erica Caple, 'Ruptures, Rights, and Repair: The Political Economy of Trauma in Haiti', *Social Science & Medicine*, 70 (2010), 106–13

James, Erica Caple, 'Haiti, Insecurity, and the Politics of Asylum', *Medical Anthropology Quarterly*, 25 (2011), 357–76

Jeffery, Susan E., 'The Creation of Vulnerability to Natural Disasters: Case Studies from the Dominican Republic', *Disasters*, 6 (1982), 38–43

'Je t'écris Haïti', *Libération*, 19 January 2010 www.etonnantsoyageurs.com/spip.php? article4894 [accessed 12 August 2014]

Jorden, Miriam, 'Trump Administration Ends Temporary Protection for Haitians', *New York Times*, 20 November 2017 www.nytimes.com/2017/11/20/us/haitians-temporary-status.html [accessed 11 January 2018]

Joseph, Myrtho, Fahui Wang, and Lei Wang, 'GIS-based assessment of urban environmental quality in Port-au-Prince, Haiti', *Habitat International*, 41(2014), 33–40

Just Foreign Policy, *Haiti Cholera Counter* www.justforeignpolicy.org/haiti-cholera-counter [accessed 16 February 2018]

Kaisary, Philip, *The Haitian Revolution in the Literary Imagination: Radical Horizons, Conservative Constraints* (Charlottesville and London: University of Virginia Press, 2014)

Kaplan, E. Ann, *Trauma Culture: The Politics of Terror and Loss in Media and Literature* (New Brunswick, NJ: Rutgers University Press, 2005)

Kaplan, Jeffrey S., 'Young Adult Literature in the 21st Century Moving Beyond Traditional Constraints and Conventions', *The ALAN Review*, 32 (2005), 11–18

Katz, Jonathan M., *The Big Truck That Went By: How the World Came to Save Haiti and Left Behind a Disaster* (New York, NY: Palgrave Macmillan, 2013)

Kelman, Ilan, ed., 'Understanding Vulnerability to Understand Disasters' Version 3, 19 January 2009 (Version 1 was 2 September 2007) www.islandvulnerability.org/docs/vulnres.pdf [accessed 12 February 2018]

Kelman, Ilan, 'How resilient is resilience?', Urban Resilience Research Network, 20 April 2016, www.urbanresilienceresearch.net/2016/04/20/how-resilient-is-resilience/ [accessed 2 February 2018]

Kelman, Ilan, J. C. Gaillard, James Lewis, and Jessica Mercer, 'Learning from the history of disaster vulnerability and resilience research and practice for climate change', *Natural Hazards*, 82 (2016), S129–S143

Kelman, Ilan, J. C. Gaillard, Jessica Mercer, 'Climate Change's Role in Disaster Risk Reduction's Future: Beyond Vulnerability and Resilience', *International Journal of Disaster Risk Science*, 6 (2015), 21–7

Kelman, Ilan, James Lewis, J.C. Gaillard, and Jessica Mercer, 'Participatory Action Research for Dealing with Disasters on Islands', *Island Studies Journal*, 6 (2011), 59–86

Kelman, Ilan, Jessica Mercer, and Jennifer West, 'Combining Different Knowledges: Community-based Climate Change Adaptation in Small Island Developing States', *Participatory Learning and Action Notes*, 60 (2009), 41–53

Kirksey, Eben, 'Hope', *Environmental Humanities*, 5 (2014), 295–300

Klein, Kerwin Lee, 'On the Emergence of Memory in Historical Discourse', *Representations: Special Issue: Grounds for Remembering*, 69 (2000), 127–50

Koening, Harold G., *In the Wake of Disaster: Religious Responses to Terrorism and Catastrophe* (Philadelphia, PA: Templeton Foundation Press, 2006)

Laferrière, Dany, *Tout bouge autour de moi* (Paris: Éditions Grassset & Fasquelle, 2011)

Laferrière, Dany, *Tout bouge autour de moi* (Montréal: Mémoire d'encrier, 2011)

Laferrière, Dany, *The World is Moving Around Me: A Memoir of the Haiti Earthquake*, trans. by David Homel (Vancouver: Arsenal Pulp Press, 2013)

Lahens, Yanick, *Failles* (Paris: Sabine Wespieser, 2010)

Lake, Nick, *In Darkness* (London: Bloomsbury, 2012)

Lathan, Don, 'The Cultural Work of Magical Realism in Three Young Adult Novels', *Children's Literature in Education*, 38 (2007), pp. 59–70

Lauro, Sarah Juliet, *The Transatlantic Zombie: Slavery, Rebellion, and Living Death* (New Brunswick: Rutgers University Press, 2015)

Law, Jonathan and Elizabeth A. Martin, 'restitutio in integrum', in *A Dictionary of Law Online* (Oxford University Press, 2009) www.oxfordreference.com/10.1093/acref/9780199551248.001.0001/acref-9780199551248-e-3401 [accessed 24 February 2015]

Leak, Andrew, 'A Vain Fascination: Writing from and about Haiti after the Earthquake', *Bulletin of Latin American Research*, 32 (2013) 394–406

Le Goff, Jacques, *History and Memory*, trans. by Steven Rendall and Elizabeth Claman (New York, NY: Columbia University Press, 1992)

Lejeune, Philippe, *Le Pact autobiographique* (Paris: Éditions du Seuil 1975)

Lewis, James, 'Some Aspects of Disaster Research', *Disasters*, 1 (1977), 241–4

Lewis, James, 'Natural Hazard Reduction', *Environment*, 30 (1988), 3–4

Lewis, James, *Development in Disaster-Prone Places: Studies of vulnerability*, (London: Intermediate Technology Publications, 1999)

'Living Lexicon', *Environmentalhumanities.org* http://environmentalhumanities.org/lexicon/ [accessed 10 February 2018]

Loughrey, Bryan, *The Pastoral Mode: A Casebook* (London: Macmillan, 1984)

Luckhurst, Roger, *The Trauma Question* (London: Routledge, 2008)

Luckhurst, Roger, *Zombies: A Cultural History* (London: Reaktion Book, 2015)

Lundahl, Mats, 'History as an Obstacle to Change: The Case of Haiti', *Journal of Inter-american Studies and World Affairs*, 31 (1989), 1–21

Lundy, Garvey, 'Price-Mars, Jean', in *The Oxford Encyclopedia of African Thought*, ed. by Abiola Irele and Biodun Jeyifo (New York, NY: Oxford University Press, 2010), pp. 258–60

Lundy, Garvey, 'The Haiti Earthquake of 2010: The Politics of a Natural Disaster', *Journal of Black Studies*, 42 (2011), 127–30

Lyotard, Jean-François, *Heidegger and 'the Jews'*, trans. by Andreas Michel and Mark S. Roberts (Minneapolis, MN. and London: University of Minnesota Press, 1997)

Macnaghten, Phil and John Urry, *Contested Natures* (London: SAGE, 1998)

Malpas, Jeff, *Place and Experience: A Philosophical Topography* (Cambridge: Cambridge University Press, 1999)

Manyena, Siambabala Bernard, 'The concept of resilience revisited', *Disasters*, 30 (2006), 433–50

Marinelli, Peter V., *Pastoral* (London: Methuen, 1971)

Martin, Lionel-Édouard, *Le Tremblement: Haïti, 12 janvier 2010* (Paris: Arléa, 2010)

Marquez Stathis, Sandra, *Rubble: The Search For a Haitian Boy* (Guilford, CT: Lyons Press, 2012)

Marx, Leo, *The Machine in the Garden: Technology and the Pastoral Ideal in America* (Oxford: Oxford University Press, 2000)

Massumi, Brian, 'Everywhere You Want to Be: Introduction to Fear', in *The Politics of Everyday Fear*, ed. by Brian Massumi (Minneapolis, MN: University of Minnesota Press, 1993), pp. 3–38

M'bayo, Tamba E., 'Dahomey', in *The Oxford Encyclopedia of African Thought*, ed. by Abiola Irele and Biodun Jeyifo (New York, NY: Oxford University Press, 2010), pp. 269–71

McAlister, Elizabeth, 'Slaves, Cannibals, and Infected Hyper-Whites: The Race and Religion of Zombies', *Anthropological Quarterly*, 85 (2012), 457–86

McAlister, Elizabeth, 'Humanitarian Adhocracy, Transnational New Apostolic Missions, and Evangelical Anti-Dependency in a Haitian Refugee Camp', *Nova Religio: The Journal of Alternative and Emergent Religions*, 16 (2013), 11–34

McCarthy Brown, Karen, 'Alourdes: A Case Study of Moral Leadership in Haitian Vodou', in *Saints and Virtues*, ed. by John Stratton Hawley (Berkeley: University of California Press, 1987), pp. 144–67

McCarthy Brown, Karen, 'Afro-Caribbean Spirituality: A Haitian Case Study', in *Vodou in Haitian Life and Culture: Invisible Powers*, ed. by Claudine Michel and Patrick Bellegarde-Smith (New York, NY: Palgrave Macmillan, 2006), pp. 1–27

McIntosh, Shawn and Marc Leverette, *Zombie Culture: Autopsies of the Living Dead* (Lanham, MD: Scarecrow Press, 2008)

McLeod, John, ed., *The Routledge Companion to Postcolonial Studies* (London: Routledge, 2007)

McLeod, John, 'Introduction: Postcolonial Environments', *The Journal of Commonwealth Literature*, 51 (2016), 192–5

McSweeney, Joyelle, 'Poetics, Revelations, and Catastrophes: an Interview with Kamau Brathwaite', *Rain Taxi*, Fall (2005), www.raintaxi.com/poetics-revelations-and-catastrophes-an-interview-with-kamau-brathwaite/ [accessed 26 September 2015]

Mehta, Brinda, *Notions of Identity, Diaspora, and Gender in Caribbean Women's Writing* (New York: Palgrave Macmillan, 2009)

'memento, n.', in *Oxford English Dictionary Online* (Oxford University Press, 2015) www.oed.com/view/Entry/116329?redirectedFrom=memento [accessed 10 March 2015]

Merklen, Denis, *Management of Social Transformations MOST: Urban Development Projects: Neighbourhood, State and NGOs, Final Evaluation of the MOST Cities Project* (UNESCO, October 2000) www.unesco.org/most/dp54merklen.pdf [accessed 29 December 2014]

Mertens, Bram, ' "Hope, Yes, But Not For Us": Messianism and Redemption in the Work of Walter Benjamin', in *Messianism, Apocalypse and Redemption in 20th-Century German Thought*, ed. by Wayne Cristaudo and Wendy Baker (Adelaide: ATF Press, 2006), pp. 63–79

Michel, Claudine, '*Kalfou Danje*: Situating Haitian Studies and My Own Journey within It', in *The Haiti Exception: Anthropology and the Predicament of Narrative*, ed. by Alessandra Benedicty-Kokken, Kaiama L. Glover, Mark Schuller, and Jhon Picard Byron, (Liverpool: Liverpool University Press, 2016), pp. 193–208

Michel, Claudine and Patrick Bellegarde-Smith, '*Koko Na Pa De Pwèl: Marasa* Reflection on Gede', in *In Extremis: Death and Life in 21st-Century Haitian Art* (Los Angeles: Fowler Museum at UCLA, 2012), pp. 13–16

Middleton, David and Steven D. Brown, *The Social Psychology of Experience Studies in Remembering and Forgetting* (London: SAGE, 2005)

Moles, Peter and Nicholas Terry, 'zombie', in *The Handbook of International Financial Terms Online* (Oxford University Press, 1997) www.oxfordreference.com.wam.leeds.ac.uk/view/10.1093/acref/9780198294818.001.0001/acref-9780198294818-e-8501 [accessed 12 May 2014]

Montilus, Guérin C., 'Vodun and Social Transformation in the African Diasporic Experience: The Concept of Personhood in Haitin Vodun Religion', in *Haitian Vodou: Spirit, Myth, and Reality* ed. by Patrick Bellegarde-Smith and Claudine Michel (Bloomington: Indiana University Press, 2006), pp. 1–7

Morisseau-Leroy, Felix, 'Boat People', *Pen.org*, 10 January 2013, https://pen.org/tourist-and-boat-people/ [accessed 28 November 2017]

Moses, Peter and John Moses, 'Haiti and the Acquired Immunodeficiency Syndrome', *Annals of Internal Medicine*, 99 (1983), 565–74

'muck', in *Oxford English Dictionary Online* (Oxford University Press, 2015) www.oed.com/view/Entry/123164?rskey=5XwKKH&result=1&isAdvanced=false [accessed 17 June 2015]

Muggah, Robert, 'The Perils of Changing Donor Priorities in Fragile States: The Case of Haiti', in *Exporting Good Governance: Temptations and Challenges in Canada's Aid Programme* ed. by Jennifer Welsh and Ngaire Woods (Waterloo: Centre for International Governance Innovation and Wilfred Laurier University Press, 2008), pp. 169–203

Munro, Martin, *Exile and Post-1946 Haitian Literature: Alexis, Depestre, Ollivier, Laferrière, Danticat* (Liverpool: Liverpool University Press, 2007)

Munro, Martin, *Haiti Rising: Haitian History, Culture and the Earthquake of 2010* (Liverpool: Liverpool University Press, 2010)

Munro, Martin, *Writing on the Fault Line: Haitian Literature and the Earthquake of 2010* (Liverpool: Liverpool University Press, 2014)

Munro, Martin, *Tropical Apocalypse: Haiti and the Caribbean End Times* (Charlottesville: University of Virginia Press, 2015)

Naipaul, V. S., *The Middle Passage* (London: André Deutsch, 1962)

Nesbitt, Nick, *Universal Emancipation: The Haitian Revolution and the Radical Enlightenment* (Charlottesville, VA: University of Virginia Press, 2008)

Nesbitt, Nick, 'Alter-Rights: Haiti and the Singularization of Universal Human Rights, 1804–2004', *International Journal of Francophone Studies*, 12 (2009), 93–108

'Nick Lake', *Unitedagents.co.uk*, www.unitedagents.co.uk/nick-lake [accessed 17 February 2018]

Nietzsche, Friedrich, *Untimely Meditations*, trans. by R. J. Hollingdale (Cambridge: Cambridge University Press, 1997)

Nikolajeva, Maria, 'Beyond the Grammar of Story, or How Can Children's Literature Criticism Benefit from Narrative Theory?', *Children's Literature Association Quarterly*, 28 (2003), 5–16

Nixon, Rob, *London Calling: V. S. Naipul, Postcolonial Mandarin* (New York, NY: Oxford University Press, 1992)

Nixon, Rob, *Slow Violence and the Environmentalism of the Poor* (Cambridge, MA: Harvard University Press, 2011)

Noll, Mark A., *American Evangelical Christianity: An Introduction* (Oxford and Malden, MA: Blackwell Publishers, 2001)

O'Grady, Kari A., Deborah G. Rollison, Timothy S. Hanna, Heidi Schreiber-Pan, and Manuel A. Ruiz, 'Earthquake in Haiti: Relationship with the Sacred in Times of Trauma', *Journal of Psychology & Theology*, 40 (2012), 289–301

O'Keefe, Phil, Ken Westgate, Ben Wisner, 'Toward an Explanation of Disaster Proneness'', Occasional Paper No. 1. Disaster Research Unit, University of Bradford (1975), in Ben Wisner, 'Disaster Vulnerability: Scale, Power and Daily Life', *GeoJournal*, 30 (1993), 127–40 (128)

Oliver, Kendrick, 'How to be (the Author of) Born Again: Charles Colson and the Writing of Conversion in the Age of Evangelicalism', *Religions*, 5 (2014), 886–911

Oliver-Smith, Anthony, 'Haiti's 500-Year Earthquake', in *Tectonic Shifts: Haiti Since the Earthquake*, ed. by Mark Schuller and Pablo Morales (Sterling, VA: Stylus Publishing 2012), pp. 18–26

Olwig, Karen Fog, *Caribbean Journeys: an Ethnography of Migration and Home in Three Family Networks* (Durham and London: Duke University Press, 2007)

Onega, Susan and José Ángel García Landa, *Narratology: An Introduction* (London: Longman, 1996)

Open for Business', *The Economist*, 7 January 2012, www.economist.com/node/21542407 [accessed 19 August 2013]

Orton, James Douglas and Kari A. O'Grady, 'Cosmology Episodes: A Reconceptualization', *Journal of Management, Spirituality & Religion*, 13 (2016), 226–45

Osborne, Peter and Matthew Charles, 'Walter Benjamin', in *The Stanford Encyclopedia of Philosophy Online*, ed. by Edward N. Zalta (2012) http://plato.stanford.edu/archives/win2012/entries/benjamin/ [accessed 20 April 2013]

Oxford English Dictionary Online (2012), www.oed.com

Padgett, Tim, 'The Failed State That Keeps Failing: Quake-Ravaged Haiti Still without a Government', *Time Online*, 10 September 2011 http://world.time.com/2011/09/10/

the-failed-state-that-keeps-failing-quake-ravaged-haiti-still-without-a-government [accessed 15 July 2013]

Padgett, Tim, 'Chile and Haiti: A Tale of Two Earthquakes', *Time*, 1 March 2010, http:// content.time.com/time/world/article/0,8599,1968576,00.html [accessed 13 February 2018]

Paget, Henry, 'The Caribbean Plantation: Its Contemporary Significance', in *Sugar, Slavery, and Society: Perspectives on the Caribbean, India, the Mascarenes, and the United States*, ed. by Bernard Moitt (Gainesville: University Press of Florida, 2004), pp. 157–87

'palimpsest, n. and adj.', *OED Online*, Oxford University Press (2018) www.oed.com/ view/Entry/136319 [accessed 22 February 2018]

Pallardy, Richard, 'Haiti earthquake of 2010', *Encyclopædia Britannica*, 28 June 2017, www.britannica.com/event/Haiti-earthquake-of-2010 [accessed 12 October 2017]

Paravisini-Gebert, Lizabeth, '"All Misfortune Comes from the Cut Trees": Marie Chauvet's Environmental Imagination', *Yale French Studies*, 128 (2015), 74–91

Park, Chris, 'environment', in *Oxford Dictionary of Environment and Conservation Online* (Oxford University Press, 2013) www.oxfordreference.com/view/10.1093/ acref/9780199641666.001.0001/acref-9780199641666-e-2563 [accessed 9 March 2015]

'pathos, n.', in *Oxford English Dictionary Online* (Oxford University Press, 2013) www. oed.com/view/Entry/138808?redirectedFrom=pathos [accessed 1 August 2013]

Patterson, Annabel, *Pastoral and Ideology: From Virgil to Valéry* (Berkeley, CA: University of California Press, 1987)

Paul VI, 'Constitution on the Sacred Liturgy Sacrosanctum Concilium', (1963) www. vatican.va/archive/hist_councils/ii_vatican_council/documents/vat-ii_const_1963 1204_sacrosanctum-concilium_en.html (para. 102 of 130) [accessed 08 Deceomber 2012]

Pelling, Mark, *The Vulnerability of Cities: Natural Disasters and Social Resilience* (London and Sterling, VA: Earthscan Publications, 2003)

Phelan, James, 'Narrative Technique', in *The Encyclopedia of The Novel*, ed. by Peter Melville Logan (Oxford: Wiley Blackwell, 2011), pp. 549–53 (p. 550).

Plant, Roger, *Sugar and Modern Slavery: A Tale of Two Countries* (London and Atlantic Highlands, NJ: Zed Books, 1987)

Polyné, Millery, ed., *The Idea of Haiti: Rethinking Crisis and Development* (Minneapolis and London: University of Minnesota Press, 2013)

Polyné, Millery, 'To Make Visible the Invisible Epistemological Order: Haiti, Singularity, and Newness', in *The Idea of Haiti: Rethinking Crisis and Development*, ed. by Millery Polyné (Minneapolis and London: University of Minnesota Press, 2013), pp. xi–xxvii.

Popham, Peter, 'Haiti: The Graveyard of Hope ', *Independent*, 13 January 2013 www. independent.co.uk/news/world/americas/haiti-the-graveyard-of-hope-8449775.html [accessed 02 May 2013]

Popkin, Jeremy D., 'Life in the Ruins: Personal Narratives and the Haitian Earthquake of 2010', *L'Esprit Créateur*, 56 (2016), 101–15

Potter, Amy E., 'Voodoo, Zombies, and Mermaids: US Newspaper Coverage of Haiti', *The Geographical Review*, 99 (2009), 208–30

Pressley-Sanon, Toni, 'Haiti: Witnessing as Revolutionary Praxis in Raoul Peck's Films', *Black Camera*, 5 (2013), 34–55

Pressley-Sanon, Toni, 'One Plus One Equals Three: Marasa Consciousness, the Lwa, and Three Stories', *Research in African Literatures*, 44 (2013), 118–37

Quayson, Ato, *Postcolonialism: Theory, Practice or Process?* (Cambridge: Polity, 2000)

Quinby, Lee, 'The Subject of Memoirs: The Woman Warrior's Technology of Ideographic Selfhood', in *De/Colonizing the Subject: The Politics of Gender in Women's Autobiography*, ed. by Sidonie Smith and Julia Watson (Minneapolis, MN: University of Minnesota Press, 1992), pp. 297–321

Rahill, Guitele J., N. Emel Ganapati, J. Calixte Clérismé, and Anuradha Mukherji, 'Shelter recovery in urban Haiti after the earthquake: the dual role of social capital', *Disasters*, 38 (S1) (2014), S73–S93

Rahill, Guitele, J., Emel Ganapati, Manisha Joshi, Brittany Bristol, Amanda Molé, Arielle Jean-Pierre, Ariele Dionne, and Michele Benavides, 'In Their Own Words: Resilience among Haitian Survivors of the 2010 Earthquake', *Journal of Health Care for the Poor and Underserved*, 27 (2016) 580–603

Rak, Julie, 'Are Memoirs Autobiography? A Consideration of Genre and Public Identity', *Genre*, 36 (2004), 305–26

Ramsey, Kate, *The Spirits and the Law: Vodou and Power in Haiti* (Chicago and London: University of Chicago Press, 2011)

'reconstruire', in *Le Trésor de la Langue Française Informatisé* (2012) www.cnrtl.fr/definition/reconstruire [accessed 08 December 2012]

Reid, Julian, 'The Disastrous and Politically Debased Subject of Resilience', *Development Dialogue: The End of the Development-Security Nexus? The Rise of Global Disaster Management*, 58 (2012), 67–79

'réplique', in *Reverso: Collins Online English-French Dictionary* (HarperCollins Publishers, 2012) http://dictionary.reverso.net/french-english/replique [accessed 1 June 2017]

'requiem', in *World Encyclopaedia Online* (Philip's, 2012) www.oxfordreference.com.wam.leeds.ac.uk/view/10.1093/acref/9780199546091.001.0001/acref-9780199546091-e-9811?rskey=NFhIGO&result=1&q=requiem [accessed 01 February 2013]

'restitutio in integrum, n.', in *Oxford English Dictionary Online* (Oxford University Press, 2015) www.oed.com/view/Entry/163965?redirectedFrom=restitutio+in+integrum [accessed 24 February 2015]

Reuben, Anthony, 'Why did fewer die in Chile's earthquake than in Haiti's?', *BBC News*, 1 March 2010, http://news.bbc.co.uk/2/hi/americas/8543324.stm [accessed 13 February 2018]

Rey, Terry, 'Catholicism and Human Rights in Haiti: Past, Present, and Future', *Religion & Human Rights*, 1 (2006), 229–48

Rey, Terry and Karen Richman, 'The Somatics of Syncretism: Tying Body and Soul in Haitian Religion', *Studies in Religion/Sciences Religieuses*, 39 (3), 379–403

Richman, Karen, 'Religion at the Epicenter: Agency and Affiliation in Léogâne After the Earthquake', *Studies in Religion/Sciences Religieuses*, 41 (2012), 148–65

Ringgren, Helmer, 'Messianism: An Overview', in *Encyclopedia of Religion*, ed. by Lindsay Jones (Detroit, MI: Macmillan Reference, 2005), pp. 5972–4

Rofles, Manfred, 'Poverty tourism: theoretical reflections and empirical findings regarding an extraordinary form of tourism', *GeoJournal*, 75 (2010), 421–42

Roxburgh, Stephen, 'The Art of the Young Adult Novel. Keynote Adress: ALAN Workshop, Indianapolis, IN., November 20, 2004', *The ALAN Review*, 32 (2005), 4–10

Rowland, Christopher and John Barton, eds., *Apocalyptic in History and Tradition* (Sheffield: Sheffield Academic Press, 2002)

'ruin, n.', in *Oxford English Dictionary Online* (Oxford University Press, 2013) www.oed.com/view/Entry/168689?rskey=rk5JU5&result=1&isAdvanced=false [accessed 12 March 2013]

'ruin, v.', in *Oxford English Dictionary Online* (Oxford University Press, 2013) www.
   oed.com/view/Entry/168690?rskey=rk5JU5&result=2&isAdvanced=false
   [accessed 12 March 2013]

'sacrifice', in *Oxford English Dictionary Online* (Oxford University Press, 2012) www.
   oed.com/view/Entry/169571?rskey=4fvYPa&result=1&isAdvanced=false#eid
   [accessed 05 November 2012]

Saghafi, Kas, 'Dying Alive', *Mosaic: a journal for the interdisciplinary study of literature*, 48 (2015), 15–26

Saint-Éloi, Rodney, *Haïti, kenbe la! 35 secondes et mon pays à reconstruire* (Neuilly-sur-
   Seine: Éditions Michel Lafon, 2010)

Sales, Roger *English Literature in History 1780–1830: Pastoral and Politics* (London:
   Hutchinson, 1983)

Schlunke, Katrina, 'Burnt houses and the haunted home', in *Housing and Home
   Unbound: Intersections in economics, environment and politics in Australia*, ed. by
   Nicole Cook, Aidan Davison, Louise Crabtree (New York: Routledge), pp. 218–31

Schmid, Wolf, 'Narratee', in *The Living Handbook of Narratology Online*, ed.by Peter
   Hühn, Jan Christoph Meister, John Pier, and Wolf Schmid (Hamburg: Hamburg
   University, 2013) www.lhn.uni-hamburg.de/article/narratee [accessed 19 August 2015]

Schuller Mark, and Pablo Morales, 'Haiti's Vulnerability to Disasters', in *Tectonic Shifts:
   Haiti Since the Earthquake*, ed. by Mark Schuller and Pablo Morales (Sterling, VA:
   Stylus Publishing 2012), pp. 11–13.

Scott, David, *Conscripts of Modernity: The Tragedy of Colonial Enlightenment* (Durham,
   NC. and London: Duke University Press, 2004)

Schiller, Nina Glick, 'Forward: Locality, Globality and the Popularization of a Diasporic
   Consciousness: Learning from the Haitian Case', in *Geographies of the Haitian
   Diaspora*, ed. by Regine O. Jackson (New York: Routledge, 2011), pp. xxi-xxix

Schininà, Guglielmo, Mazen Aboul Hosn, Amal Ataya, Kety Dieuveut, and Marie-Ade'le
   Salem, 'Psychosocial Response to the Haiti Earthquake: the Experiences of International Organization for Migration', *Intervention*, 8 (2010), 158–64

Schuster, Joshua, 'How to Write the Disaster', *The Minnesota Review*, 83 (2014), 163–71

Sepinwall Alyssa Goldstein, *Haitian History: New Perspectives* (New York, NY:
   Routledge, 2013)

Shapiro, Stephen A., 'The Dark Continent of Literature: Autobiography', *Comparative
   Literature Studies*, 5 (1968), 421–54

Sharpe, Christina, *In the Wake: On Blackness and Being* (Durham and London: Duke
   University Press, 2016)

Sheinin, David, 'The Caribbean and the Cold War: Between Reform and Revolution', in *The
   Caribbean: A History of the Region and Its Peoples*, ed. by Stephan Palmié and Francisco
   A. Scarano (Chicago, IL. and London: University of Chicago Press, 2011), pp. 491–505

Sheller, Mimi, *Aluminum Dreams: The Making of Light Modernity* (Cambridge, MA. and
   London: MIT Press, 2014)

'sheol, n.', in *Oxford English Dictionary Online* (Oxford University Press, 2014) www.
   oed.com/view/Entry/177962?redirectedFrom=sheol [accessed 3 March 2014]

Skrimshire, Stefan 'How Should We Think About the Future', in *Future Ethics: Climate
   Change and Apocalyptic Imagination*, ed. by Stefan Skrimshire (London: Continuum,
   2010), pp. 1–11

Smith, Katherine, 'Genealogies of Gede', in *In Extremis: Death and Life in 21st-Century
   Haitian Art*, ed.by Donald J. Cosentino (Los Angeles: Fowler Museum at UCLA,
   2012), pp. 85–99

Smith, Matthew J., ' "A Tale of Two Tragedies: Remembering and Forgetting Kingston and Port-au-Prince", Shaking Up the World? Global Effects of Haitian Tremors: 1791, 2010', (Aarhus University, Denmark; 10–12 August 2017)

Sontag, Susan, *On Photography* (London: Penguin Books, 2002)

Sotter, Anna O. and Sean P. Connors, 'Beyond Relevance to Literary Merit: Young Adult Literature as "Literature" ', *The ALAN Review*, 37 (2009), 62–7

Southwick, Steven M., George A. Bonanno, Ann S. Masten, Catherine Panter-Brick, and Rachel Yehuda, 'Resilience Definitions, Theory, and Challenges: Interdisciplinary Perspectives', *European Journal of Psychotraumatology*, 5 (2014) 10.3402/ejpt.v5.25338 [accessed 25 September 2017]

'souvenir', in *Oxford English Dictionary Online* (Oxford University Press, 2013) www.oed.com/view/Entry/185321?rskey=0KQdXq&result=1&isAdvanced=false#ei [accessed 12 April 2013]

Spengemann, William C. and L. R. Lundquist, 'Autobiography and the American Myth', *American Quarterly*, 17 (1965), 501–19

Steinbrink, Malte, Fabian Frenzil, and Ko Koens, 'Development and globalization of a new trend in tourism', in *Slum Tourism: Poverty, Power and Ethics*, ed. by Fabian Frenzel, Malte Steinbrink, and Ko Koens (New York, NY: Routledge, 2012), pp. 1–18

Sternberg, Esther M., 'Walter B. Cannon and " ' Voodoo' Death": A Perspective From 60 Years On', *American Journal of Public Health*, 92 (2002), 1564–6

Stoler, Ann Laura, 'Imperial Debris: Reflections on Ruins and Ruination', *Cultural Anthropology*, 23 (2008), 191–221

Stoler, Ann Laura, *Duress: Imperial Durabilities in Our Times* (Durham: Duke University Press, 2016)

Swanson, Lucy, 'Zombie Nation? The Horde, Social Uprisings, and National Narratives', *Cincinnati Romance Review*, 34 (2012), 13–33

'témoignagne', in *Reverso: Collins Online English-French Dictionary* (HarperCollins Publishers, 2012) http://dictionary.reverso.net/french-english/t%C3%A9moignage [accessed 26 May 2015]

Terrier, Monique, Anne Bialkowski, C Prépetit, Aude Nachbauer, and Y. F. Joseph, 'Revision of the Geological Context of the Port-Au-Prince Metropolitan Area', Haiti: Implications for Slope Failures and Seismic Hazard Assessment', *Natural Hazards and Earth System Sciences*, 14 (2014), 2577–87

Todorov, Tzvetan, *The Conquest of America: the Question of the Other*, trans. by Richard Howard (New York, NY: HarperPerennial, 1992).

Tobin, Kathleen A., 'Population Density and Housing in Port-au-Prince: Historical Construction of Vulnerability', *Journal of Urban History*, 39 (2013), 1045–61

'tremblement', in *Le Trésor de la Langue Française Informatisé* (2012) www.cnrtl.fr/definition/tremblement [accessed 08 December 2012]

'Trojan horse', in *Britannica Online* (2013) www.britannica.com/EBchecked/topic/606 297/Trojan-horse [accessed 13 August 2013]

Trouillot, Michel-Rolph, *Silencing the Past: Power and the Production of History* (Boston, MA: Beacon Press, 1995)

Trouillot, Evelyne, 'Eternity Lasted Less than Sixty Seconds … ', in *Haiti Rising: Haitian History, Culture and the Earthquake of 2010*, ed. by Martin Munro (Liverpool: Liverpool University Press, 2010), pp. 55–9

Turner, Billie L., Roger E. Kasperson, Pamela A. Matson, James J. McCarthy, Robert W. Corell, Lindsey Christensen, Noelle Eckley, Jeanne X. Kasperson, Amy Luers, Marybeth L. Martello, Colin Polskya, Alexander Pulsipher, and Andrew Schiller, 'A

Framework for Vulnerability Analysis in Sustainability Science', *Proceedings of the National Academy of Sciences*, 100 (2003), 8074–9

Ulysse, Gina Athena, *Why Haiti Needs New Narratives: A Post-Quake Chronicle* (Middletown, CT: Wesleyan University Press, 2015)

Ulysse, Gina Athena, 'Introduction', *e-misférica: Caribbean Rasanblaj* http://hemisphericinstitute.org/hemi/en/emisferica-121-caribbean-rasanblaj/e-121-introduction [accessed 18 December 2017]

'UN Radio: Poverty is behind Haiti's vulnerability to natural disasters' [interview], *Preventionweb.net* www.preventionweb.net/go/12306 [accessed 15 September 2017]

USGS, 'M 8.8 – offshore Bio-Bio, Chile', *Earthquake.usgs.gov* https://earthquake.usgs.gov/earthquakes/eventpage/official20100227063411530_30#executive [accessed 18 October 2017]

USGS, 'M 7.0 – South Island of New Zealand', *Earthquakes.usgs.gov* https://earthquake.usgs.gov/earthquakes/eventpage/usp000hk46#origin [accessed 18 October 2017]

USGS, National Earthquake Information, 'Magnitude 7.0 Haiti' (US Geological Survey: National Earthquake Information Center, 2013) http://earthquake.usgs.gov/earthquakes/eqinthenews/2010/us2010rja6/#summary [accessed 02 April 2013]

Vanborre, Emmanuelle Anne, 'Haïti après le tremblement de terre: le témoignage impossible et nécessaire', in *Haïti après le tremblement de terre: la forme, le rôle et le pouvoir de l'écriture* (New York, NY: Peter Lang, 2014), pp. 3–14.

van Dooren, Thom, 'Care' *Environmental Humanities*, 5 (2014), 291–4 (p. 293) https://doi.org/10.1215/22011919-3615541 [accessed 10 October 2016]

van Dooren, Thom and Deborah Bird Rose, 'Lively Ethography: Storying Animist Worlds', *Environmental Humanities*, 8 (2016): 77–94. doi: https://doi.org/10.1215/22011919-3527731 [accessed 12 February 2018]

Van Hoving, Daniël J., Lee A. Wallis, Fathima Docrat, and Shaheem De Vries, 'Haiti Disaster Tourism – A Medical Shame', *Prehospital and Disaster Medicine*, 25 (2010), 201–2

Wagner, Laura Rose, 'Salvaging', in *Haiti Rising: Haitian History, Culture and the Earthquake of 2010*, ed. by Martin Munro (Liverpool: Liverpool University Press, 2010), pp. 15–24

Wagner, Laura Rose, 'Haiti: A survivor's story', *Salon*, 2 February 2010, www.salon.com/2010/02/02/haiti_trapped_under_the_rubble/

Wagner, Laura Rose, *Hold Tight, Don't Let Go: A Novel of Haiti* (New York: Abrams, 2015) 2010/02/02/haiti_trapped_under_the_rubble/ [accessed 1 February 2017]

Walcott, Derek, 'Nobel Lecture: The Antilles: Fragments of Epic Memory', *Nobelprize.org* (Nobel Media AB: 2014)

Weheliye, Alexander G., *Habeas Viscus: Racializing Assemblages, Biopolitics, and Black Feminist Theories of the Human* (Durham: Duke University Press, 2014)

Weiner, Tim, 'Life Is Hard and Short in Haiti's Bleak Villages', *New York Times Online*, 14 March 2004 www.nytimes.com/2004/03/14/world/life-is-hard-and-short-in-haiti-s-bleak-villages.html?pagewanted=all&src=pm [accessed 15 July 2013]

Wessinger, Catherine, 'Religious Responses to the Katrina Disaster in New Orleans and the American Gulf Coast', *Journal of Religious Studies* (Japanese Association for Religious Studies), 86 (2012), 53–83

West-Pavlov, Russell, *Spaces of Fiction/Fictions of Space: Postcolonial Place and Literary Deixis* (Basingstoke: Palgrave Macmillan, 2010)

White, Gilbert F., *Natural Hazards: Local, National, Global* (Oxford: Oxford University Press, 1974), in James Lewis: 'Some Aspects of Disaster Research', *Disasters*, 1 (1977), 241–4

Winks, Christopher, 'Martin Munro *Tropical Apocalypse: Haiti and the Caribbean End Times* (Charlottesville: University of Virginia Press, 2015) [book review]', *New West Indian Guide/Nieuwe West-Indische Gids*, 91 (2017), 149

Winthrop, John, *A Modell of Christian Charity (1630)* (Collections of the Massachusetts Historical Society: Boston, 1838) 7, 31–48 https://history.hanover.edu/texts/winthmod.html [accesssed 01 October 2015]

Wisner, Ben and Ilan Kelman, 'Community Resilience to Disasters', *International Encyclopedia of the Social & Behavioral Sciences Online* (Oxford: Elsevier, 2015), pp. 354–60 https://doi.org/10.1016/B978-0-08-097086-8.28019-7 [accessed 1 February 2018]

Woolley, Dan with Jennifer Schuchmann, *Unshaken: Rising From the Ruins of Hotel Montana* (Grand Rapids, MI: Zondervan, 2011)

Zombiewalk.com http://zombiewalk.com/forum/index.php [accessed 14 May 2014]

Zanotti, Laura, 'Cacophonies of Aid, Failed State Building and NGOs in Haiti: setting the stage for disaster, envisioning the future', *Third World Quarterly*, 31 (2010), 755–71

Index

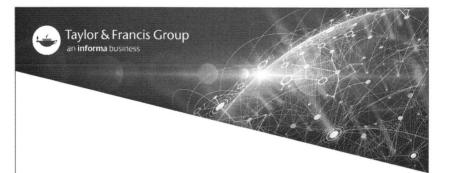

Printed and bound by CPI Group (UK) Ltd, Croydon, CR0 4YY

24/10/2024

01778282-0010